AIR QUALITY IN SOUTHEAST ASIA

STATUS, TRENDS, AND PRIORITY INTERVENTIONS ACROSS EIGHT ASEAN CITIES

MARCH 2026

ACGF
ASEAN CATALYTIC GREEN FINANCE FACILITY
An ASEAN Infrastructure Fund Initiative

ADB

6 ADB Avenue, Mandaluyong City, 1550 Metro Manila, Philippines
Tel +63 2 8632 4444; Fax +63 2 8636 2444
www.adb.org

ISBN 978-92-9277-748-7 (print); 978-92-9277-749-4 (PDF); 978-92-9277-750-0 (ebook)
Publication Stock No. TCS260121-2
DOI: http://dx.doi.org/10.22617/TCS260121-2

Notes:
In this publication, "CNY" refers to the Chinese yuan and "$" refers to United States dollars, unless otherwise stated.
ADB recognizes "China" as the People's Republic of China, "Korea" as the Republic of Korea, "Vietnam" as Viet Nam, and "Hanoi" as Ha Noi.

On the cover: Aerial view of Manila, Philippines; an electric bus on the road in Ha Noi; a windmill farm in Viet Nam; and a recycling facility in Indonesia (photos by ADB).

Contents

Tables, Figures, and Box

Tables

Figures

Box

Acknowledgments

This report is funded by the Asian Development Bank (ADB) under TA-10463: Southeast Asia Resilient and Water Secure Cities Facility, with additional support from the ASEAN Catalytic Green Finance Facility (ACGF) under TA-6983: Accelerating Climate Transitions through Green Finance in Southeast Asia.

The original technical report was developed by Clean Air Asia. ADB provided strategic direction, technical guidance, and peer review to ensure alignment with regional priorities and the objectives of the technical assistance. The Clean Air Asia team was led by Glynda Bathan-Baterina (deputy executive director) and Dang Espita Casanova (associate program director), with contributions from Everlyn Tamayo (air quality and climate change science lead), Precious Benjamin (sustainable cities lead), Enrique Cosep (air quality specialist), Carlo Bontia (senior researcher), Annette Jose (research assistant), and Ma. Loisa Tong (consultant). Additional contributions were provided by Ahn Meg Adonis (sustainable transport lead), Myron Alcanzare (senior transport researcher), Dana Lightha Babela (senior researcher), Cris Paulo Beo (researcher), Jahnea Bermejo (research assistant), Allysa Aguilar (research assistant), Andrea Leong (research assistant), and Caitlin Bernal (project intern). Support on data collection was provided by Clean Air Asia Indonesia.

The report also benefited from peer review and technical comments provided by Karma Yangzom (principal environment specialist), Alexander Nash (senior urban development specialist), Chinggay Jayme (senior environment officer), Binh Bui (climate change officer), Ma. Katherina Patdu (environment specialist), and Bethany Cosgrove (environment specialist).

The authors gratefully acknowledge the valuable input and data provided by government agencies, partner institutions, and stakeholders across the Association of Southeast Asian Nations. Special appreciation is extended to the eight cities of the case study—Bangkok, Bogor, Ha Noi, Ho Chi Minh City, Iloilo City, Jakarta, Phnom Penh, and Quezon City—for sharing their experiences and priorities in advancing air quality management.

The original technical report prepared by Clean Air Asia was edited and developed into this publication under the leadership of Sree Kartha (ACGF knowledge lead), with guidance from Scott Roberts (ACGF head), and support from Gabrielle Teodoro-Marcelino (ACGF consultant). Production coordination was managed by Ann Mushayt W. Alemania (operations officer). Copy editing was done by Corie Desuasido, design and layout by Rocilyn Laccay, proofreading by Jess Alfonso A. Macasaet, and page proof checking by Ma. Cecilia Abellar.

Abbreviations

ACGF	ASEAN Catalytic Green Finance Facility
ADB	Asian Development Bank
AGMF	ASEAN Governors and Mayors Forum
API	air pollution index
AQG	Air Quality Guidelines
AQHI	Air Quality and Health Index
AQI	air quality index
AQMt	air quality monitoring
AQMP	Air Quality Management Plan
ASEAN	Association of Southeast Asian Nations
BAU	business-as-usual
BC	black carbon
BEV	battery electric vehicle
CAAP	Clean air action plan
CCAC	Climate and Clean Air Coalition
CFP	coal-fired power plant
GDP	gross domestic product
GHG	greenhouse gas
HDV	heavy-duty vehicle
HEV	hybrid electric vehicle
LDV	light-duty vehicle
LEZ	low-emission zone
NDC	nationally determined contribution
NO_2	nitrogen dioxide
NO_X	nitrogen oxides
O_3	ground-level ozone
PHEV	plug-in hybrid electric vehicle
PSI	Pollutant Standards Index
SERD	Southeast Asia Department
SO_2	sulfur dioxide
SO_X	sulfur oxides
WHO	World Health Organization

Executive Summary

Air pollution is a major public health and environmental challenge across Southeast Asia. The *State of Global Air 2025* report estimates that air pollution contributed to 7.9 million deaths worldwide in 2023, equivalent to about one in eight deaths globally, making it the second leading risk factor for mortality after high blood pressure. The health burden is not evenly distributed across age groups. Children and older adults are among the most vulnerable, with exposure to ambient fine particulate matter ($PM_{2.5}$) linked to about 171,000 deaths globally among children under 5 years of age, while nearly 5 million deaths associated with air pollution occurred among people over the age of 70 in 2023. Earlier analysis presented in the *State of Global Air 2024* report, developed in partnership with United Nations Children's Fund (UNICEF), estimates that around 700,000 children under 5 died in 2021 due to combined exposure to ambient and household air pollution. This represents about 15% of all global deaths among children under 5, equivalent to nearly 2,000 child deaths each day. Evidence is also growing on links between air pollution and chronic disease, including findings that more than one-quarter of dementia deaths worldwide are associated with air pollution exposure.

Low- and middle-income countries bear the majority of this burden. The region covering Southeast Asia, East Asia, and Oceania accounts for about 35% of global air pollution deaths, with countries such as Indonesia and Myanmar each recording more than 100,000 deaths annually. When adjusted for population size, the Lao People's Democratic Republic and Myanmar record the highest mortality rates in the region (State of Global Air 2025). Across member states of the Association of Southeast Asian Nations (ASEAN), annual average concentrations of $PM_{2.5}$ in most urban centers exceed the World Health Organization (WHO) Air Quality Guidelines (AQG) of 5 micrograms per cubic meter ($\mu g/m^3$), exposing large populations to elevated health risks.

The impacts extend beyond health, affecting labor productivity, health care expenditures, and environmental quality. Although some cities have recorded improvements in recent years, progress has been uneven and remains insufficient to meet the WHO guideline thresholds. This highlights the need for sustained, evidence-based policy responses and underscores the scale of implementation challenges facing governments across the region.

This report provides both a regional assessment of air quality in Southeast Asia and a city-level analysis of priority actions. It covers 10 ASEAN member states—Brunei Darussalam, Cambodia, Indonesia, the Lao People's Democratic Republic, Malaysia, Myanmar, the Philippines, Singapore, Thailand, and Viet Nam—and presents updated evidence on air quality levels, monitoring systems, emission sources, and policy responses. It also features priority clean air interventions drawn from eight major ASEAN cities—Bangkok, Bogor, Ha Noi, Ho Chi Minh City, Iloilo City, Jakarta, Phnom Penh, and Quezon City—highlighting pollution drivers and practical solutions that can catalyze action.

Based on publicly available data from more than 200 cities between 2014 and 2024, the report presents the following:

- **Air quality trends.** All assessed ASEAN cities exceeded the WHO AQG for $PM_{2.5}$ in 2024, with most recording concentrations around the WHO Interim Target (IT) 2 (25 $\mu g/m^3$).
- **Monitoring capacity.** Monitoring networks are expanding but remain insufficient relative to urban population density. Public accessibility of $PM_{2.5}$ data varies across countries.
- **Emission sources.** Regional modeling indicates that residential fuel use contributes more than 25% of $PM_{2.5}$ emissions, followed by energy and industry (22%), transport (10%), and agriculture and waste (approximately 5% each). However, publicly available studies on national emissions inventories and source apportionment studies show important differences across countries. Transport is the leading source of $PM_{2.5}$ emissions in four ASEAN countries, while industry, biomass burning, and other area sources—diffuse emissions from residential, agricultural, and other small-scale activities—are the primary contributors in several others.

- **Policy measures.** Most countries have established clean air laws or national plans, and seven explicitly link air quality to public health; however, national standards generally remain less stringent than the WHO AQG levels.

City-level leadership is emerging as an important driver of progress. Case studies from the eight cities demonstrate how local governments are developing clean air strategies, which include piloting transport electrification, improving waste management systems, and testing alternatives to agricultural burning. These examples illustrate how empowered cities can advance clean air objectives while delivering climate and public health co-benefits.

Key Actions for Cleaner Air

Accelerate progress toward WHO AQG.
Several ASEAN cities have shown air quality improvements in recent years, particularly in meeting the more stringent WHO Interim Targets (ITs). However, reaching the WHO AQG will require strengthened efforts. Building on existing momentum by scaling up mitigation measures and ensuring effective implementation will be critical.

Expand and improve air quality monitoring coverage.
More extensive air quality monitoring (AQMt) is needed to support data-driven efforts to reduce air pollution and guide policy implementation, particularly in high-density urban areas and in countries where the number of monitoring stations remain limited. In addition to PM_{10} and $PM_{2.5}$, AQMt should be expanded to include pollutants such as ground-level ozone (O_3), nitrogen dioxide (NO_2), sulfur dioxide (SO_2), black carbon, elemental carbon, and ultrafine particles ($PM_{0.1}$) to provide a more comprehensive understanding of air pollution and associated health risks.

Enhance transparency and public access to air quality data.
Governments should make real-time and historical air quality data accessible through official websites, platforms, dashboards, and reports. Open data access informs the public about health advisories and supports research, policy development, and better AQMt planning. Clear documentation of monitoring station characteristics such as site location, type, and description is essential for accurate interpretation and effective use of data.

Develop and harmonize emission inventories and source apportionment studies.
Identifying specific emission sources and their relative contributions to air pollution enables more targeted interventions. Countries should develop comprehensive national and subnational emission inventories and conduct source apportionment studies to identify priority sectors and key pollution sources. Detailed local-level assessments can support integrated planning and more effective mitigation strategies. To strengthen regional cooperation, ASEAN countries should adopt harmonized protocols that improve data quality, ensure information integration, and enable coordinated responses.

Advance air quality governance at national and subnational levels.
National laws should remain responsive to emerging scientific evidence, socioeconomic changes, and technological developments. Periodic review and amendment of air quality-related legal frameworks are highly recommended to ensure that policies are aligned with evolving air quality priorities. Several cities in the region have already demonstrated strong leadership by developing and implementing local clean air action plans (CAAPs) that are tailored to their specific challenges. Strengthening engagement between national and subnational authorities can amplify air quality improvement efforts and accelerate progress.

Improve ambient air quality standards.
ASEAN countries have established national ambient air quality standards that reflect their unique conditions and priorities. As global guidelines evolve, countries should progressively align their standards with the WHO AQG. This alignment would enhance public health protection, improve accountability, and support strategies for stronger air quality improvements.

Leverage climate commitments to deliver air quality co-benefits.
ASEAN countries should explicitly integrate air quality objectives into the implementation of their nationally determined contributions. This approach promotes efficient use of resources and supports coordinated action across climate and air quality agendas.

Strengthen integration of air quality and public health management.
Most ASEAN countries have begun integrating air quality management into health planning frameworks. Building on this progress, governments should further align air quality targets with public health goals, such as exposure reduction and disease prevention, to support more proactive, health-centered decision-making.

Encourage public engagement through health-oriented air quality communication.
ASEAN countries should periodically review and enhance their air quality index classification and communication frameworks to ensure clarity, relevance, and alignment with the latest scientific evidence. Transitioning toward more health-based tools, such as the air quality and health index, can improve risk awareness and enable citizens to take informed action. These measures support short-term exposure reduction during pollution episodes and foster longer-term public participation in clean air initiatives.

Scale up financing for clean air investments.
Cities face significant gaps in financing high-impact interventions and often lack the capacity to prepare bankable proposals or access suitable financing instruments. A range of tools—grants, debt, guarantees, and results-based finance—can help address these constraints. Experience from the Beijing–Tianjin–Hebei regional air quality improvement program in the People's Republic of China demonstrates how combining multiple financial instruments can mobilize large-scale investment and achieve measurable improvements in air quality concentrations. Clearly defined air quality investment programs tailored to city needs, readiness, socioeconomic conditions, and priority interventions in key emission sectors can support scalable financing solutions.

Strengthen cross-sectoral implementation in priority emission sectors.
Across the eight cities covered in this report, action plans highlight shared priorities to address emissions from transport, household fuel use and area sources, waste management, and selected industrial activities, supported by stronger data and institutional capacity. Transport measures are featured prominently, including expansion of public transit, fleet electrification, low- or green-emission zones, and the management of high carbon-emitting vehicles. Cities also prioritize reducing open burning, improving solid waste management, and promoting cleaner production and fuel-switching, where this is relevant.

By linking regional trends with city-level solutions, this report underscores that achieving cleaner air in Southeast Asia will depend not only on national policies, but also on empowered cities, adequate financing, strong data systems, and coordinated multisector action. With targeted investment and policy alignment, ASEAN cities can significantly reduce air pollution, protect public health, enhance climate resilience, and support inclusive and sustainable development.

An electric bus on the road in Ha Noi, Viet Nam (photo by ADB).

1 Air Quality Trends

ASEAN in the Past Decade

Air quality trends across cities in the Association of Southeast Asian Nations (ASEAN) over the past decade show periods of improvement, but these have not been consistent enough to achieve the World Health Organization (WHO) Air Quality Guidelines (AQG). Analysis of particulate matter ($PM_{2.5}$) data from 220 cities across the region from 2014 to 2024 indicates fluctuations in concentration levels over time rather than a sustained decline in pollution.[1] Since 2020, a larger share of cities has recorded annual average $PM_{2.5}$ concentrations aligned with WHO Interim Target (IT) 3 (15 micrograms per cubic meter [μg/m^3]), reflecting some early gains in air quality management. The share of cities exceeding WHO IT-1 declined significantly between 2019 and 2022 but increased again in 2023 and 2024. This pattern suggests that early reductions were largely temporary and primarily driven by reduced economic and transport activity during the coronavirus disease pandemic rather than long-term structural improvements. Most cities continue to fall within WHO IT-2 (25 μg/m^3), which remains far above WHO AQG of 5 μg/m^3. This trend underscores the need for strengthened and sustained air quality interventions to achieve long-term improvements.

Figure 1: Annual $PM_{2.5}$ Levels in ASEAN Cities, 2014–2024

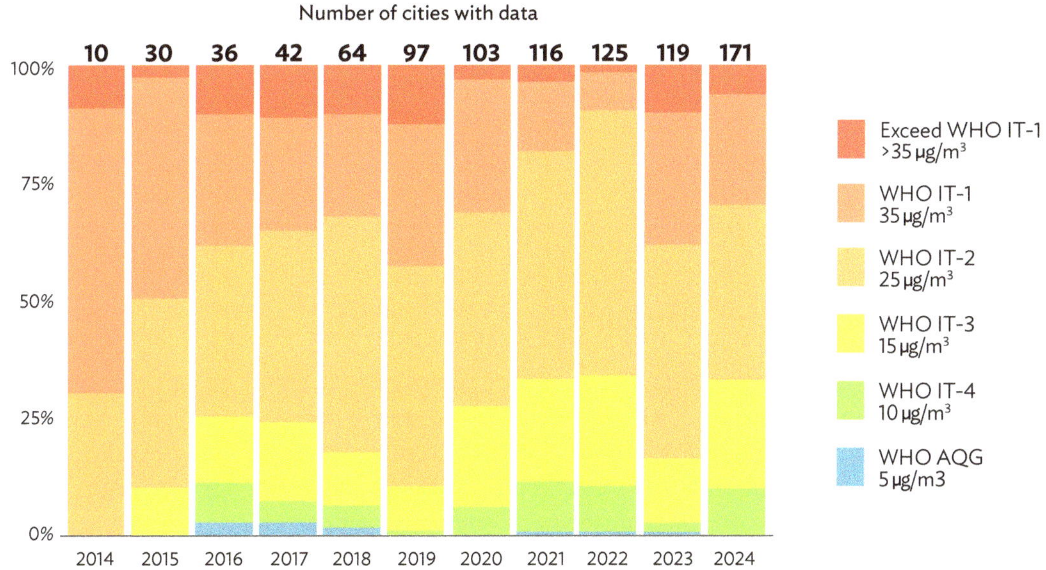

AQG = Air Quality Guidelines, ASEAN = Association of Southeast Asian Nations, IT = interim target, PM = particulate matter, WHO = World Health Organization.

Source: Clean Air Asia.

A similar pattern is observed for PM_{10}. Data compiled from 192 cities show that the share of cities exceeding WHO IT-1 (70 μg/m^3) declined markedly between 2019 and 2021, reflecting temporary improvements in air quality during the period. However, this trend reversed from 2022 to 2024, with a gradual increase in the proportion of cities exceeding WHO IT-1. Overall, most cities continue to meet only WHO IT-2 (50 μg/m^3), which remains substantially higher than WHO AQG of 15 μg/m^3. While some progress has been observed, the number of cities meeting WHO AQG remains limited, and faster progress will be required to strengthen public health protection.

[1] Data collected covers the period 2011–2024. However, only 2014–2024 is presented in this section due to limited availability of earlier data. The dataset includes both publicly accessible data and data provided by ministries upon request.

Figure 2: Annual PM_{10} Levels in ASEAN Cities, 2014-2024

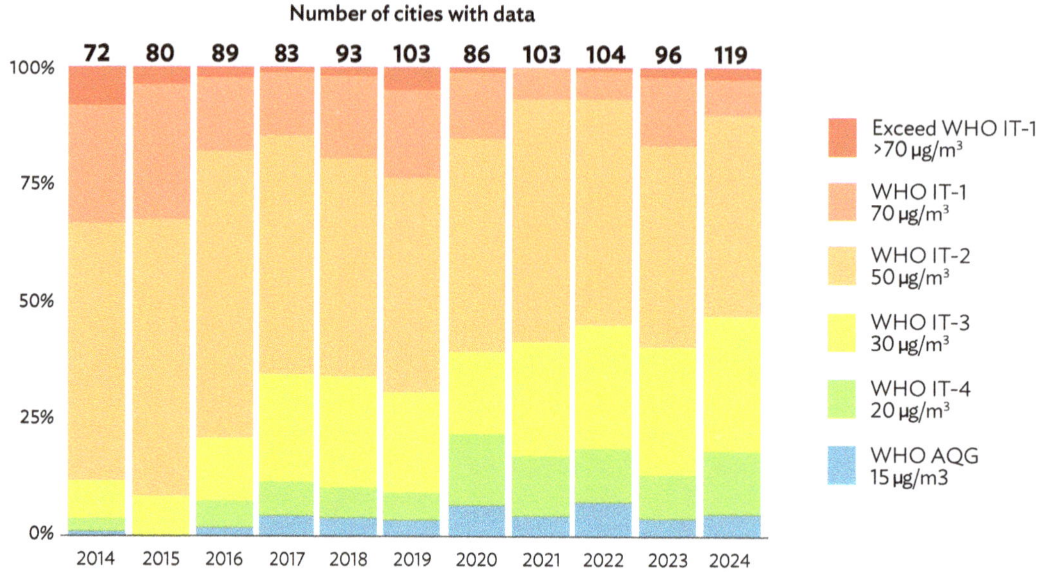

AQG = Air Quality Guidelines, ASEAN = Association of Southeast Asian Nations, IT = interim target, PM = particulate matter, WHO = World Health Organization.

Source: Clean Air Asia.

To support sustained improvements, countries need more comprehensive and consistent air quality monitoring (AQMt) as they implement additional mitigation measures. Long-term data from well-maintained monitoring stations are essential for tracking trends, assessing policy impacts, and informing future action. The number of cities with accessible $PM_{2.5}$ and PM_{10} data increased between 2014 and 2024, indicating progress in AQMt and data reporting. However, expanded monitoring coverage remains necessary to provide a more complete picture of air quality conditions not covered by existing networks.

ASEAN Cities Relative to the WHO Air Quality Guidelines

AQMt data provide critical insight into the status of air pollution relative to the WHO guideline values; however, where and how the measurements are obtained must be clearly documented to ensure accurate interpretation. Pollution levels vary significantly depending on the monitoring location. Urban and roadside sites, where human exposure is typically highest, often record elevated concentrations and should be prioritized in both monitoring design and public reporting. Background sites, including those in green spaces, are located away from major emission sources and generally show lower concentrations, reflecting broader regional air quality trends. These sites provide essential context for interpreting the data. Without transparent reporting of monitoring site characteristics, data may be misinterpreted and create a false sense of progress in cities where exposure risks remain high.

The following analysis examines the distribution of ASEAN cities relative to the WHO guideline values based on the latest available 2024 data.

$PM_{2.5}$ Levels Relative to the WHO Guidelines in 2024

Based on 2024 $PM_{2.5}$ data, all assessed ASEAN cities recorded annual average concentrations exceeding WHO AQG (5 $\mu g/m^3$), indicating continued exposure to levels associated with adverse health impacts. Among the 171 cities with available 2024 data, 17 met WHO IT-4 (10 $\mu g/m^3$), while the majority (63 of 171) fell within WHO IT-2 (25 $\mu g/m^3$). These levels remain substantially higher than WHO AQG and indicate continued exposure to elevated health risks across much of the region.

Figures 3 and 4 present the regional and country-level distribution of cities relative to WHO $PM_{2.5}$ guideline values.

Figure 3: Distribution of ASEAN Cities by 2024 $PM_{2.5}$ Levels Across WHO AQG and Interim Targets

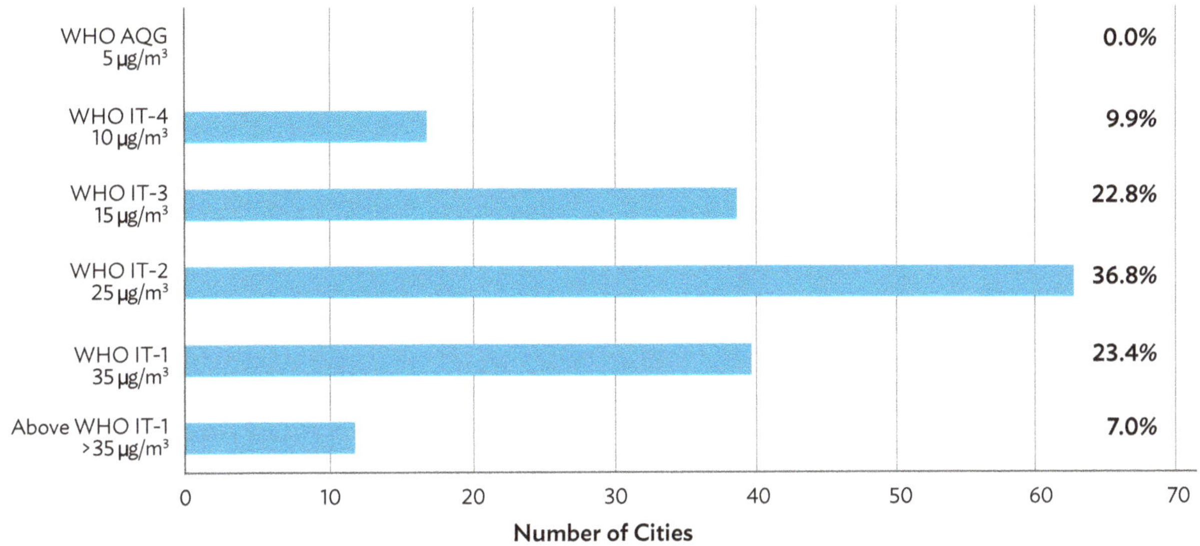

AQG = Air Quality Guidelines, ASEAN = Association of Southeast Asian Nations, IT = interim target, PM = particulate matter, WHO = World Health Organization.
Source: Clean Air Asia database on the 2024 annual average $PM_{2.5}$ concentrations in Asian cities.

Figure 4 illustrates the distribution of cities across WHO ITs, highlighting uneven progress. Only a small number met the annual $PM_{2.5}$ WHO AQG or IT guideline values. Among these, Cambodia and the Philippines accounted for more cities achieving the most stringent WHO targets. Although only two cities reported data from Brunei Darussalam, both achieved WHO IT-4 guideline. The distribution also shows that the Lower Mekong Region, particularly Thailand, has the most extensive monitoring coverage.

Figure 4: Distribution of ASEAN Cities by 2024 $PM_{2.5}$ Levels Across WHO AQG and Interim Targets, by Country

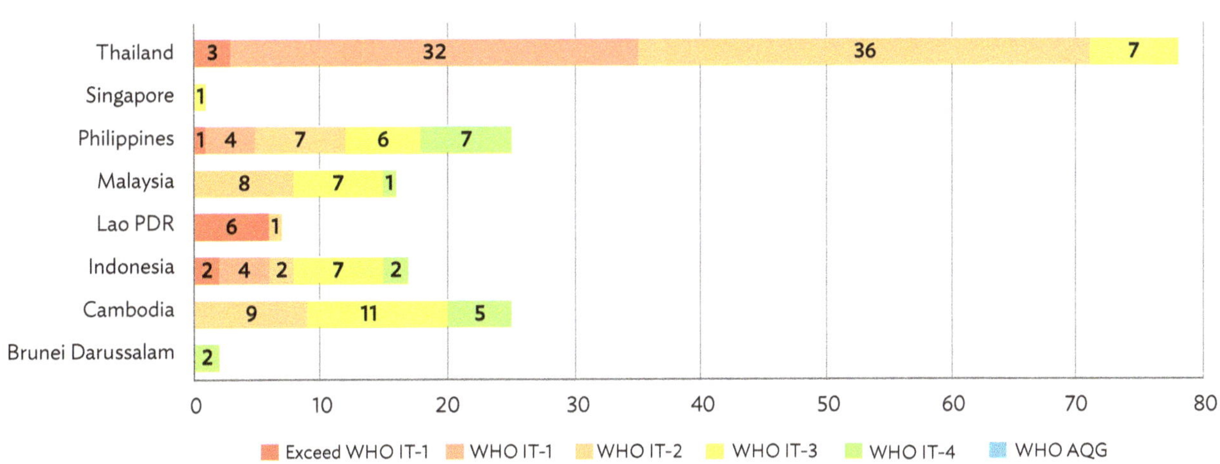

AQG = Air Quality Guidelines, ASEAN = Association of Southeast Asian Nations, IT = interim target, Lao PDR = Lao People's Democratic Republic, PM = particulate matter, WHO = World Health Organization.
Source: Clean Air Asia database on the 2024 annual average $PM_{2.5}$ concentrations in Asian cities.

PM_{10} Levels and Data Availability in 2024

PM_{10} monitoring and reporting remain more limited across ASEAN compared with $PM_{2.5}$. In 2024, only six members of ASEAN had available city-level PM_{10} data, constraining the ability to assess regional exposure comprehensively. Among the 119 cities, only 6 met WHO AQG of 15 µg/m³, indicating that the vast majority of cities (95.0%) remained above the recommended level.

On WHO IT-4 (20 µg/m³), 16 cities met this level, while 51 cities fell within WHO IT-2 (50 µg/m³). Among countries with available data, Indonesia recorded the highest number of cities meeting WHO AQG, alongside one city in Brunei Darussalam. WHO IT-4 was achieved primarily in cities in the Philippines, with additional cities in Brunei Darussalam, Indonesia, and Malaysia. Singapore, represented as a single city in the dataset, reported PM_{10} data only at the national level.

Overall, PM_{10} data availability across ASEAN remains limited. This assessment draws on publicly accessible data and information provided by ministries upon request. Nonetheless, limited public access to data remains a constraint to transparent air quality assessment and management. This gap is particularly relevant in cities affected by resuspended road dust, construction activity, and industrial emissions, where localized exposure may be significant. Strengthening monitoring coverage and improving public data access would support more evidence-based air quality planning and enhance public accountability.

Figures 5 and 6 illustrate the distribution of ASEAN cities relative to the WHO PM_{10} guideline values at the regional and country levels.

Figure 5: Distribution of ASEAN Cities by 2024 PM_{10} Levels Across WHO AQG and Interim Targets

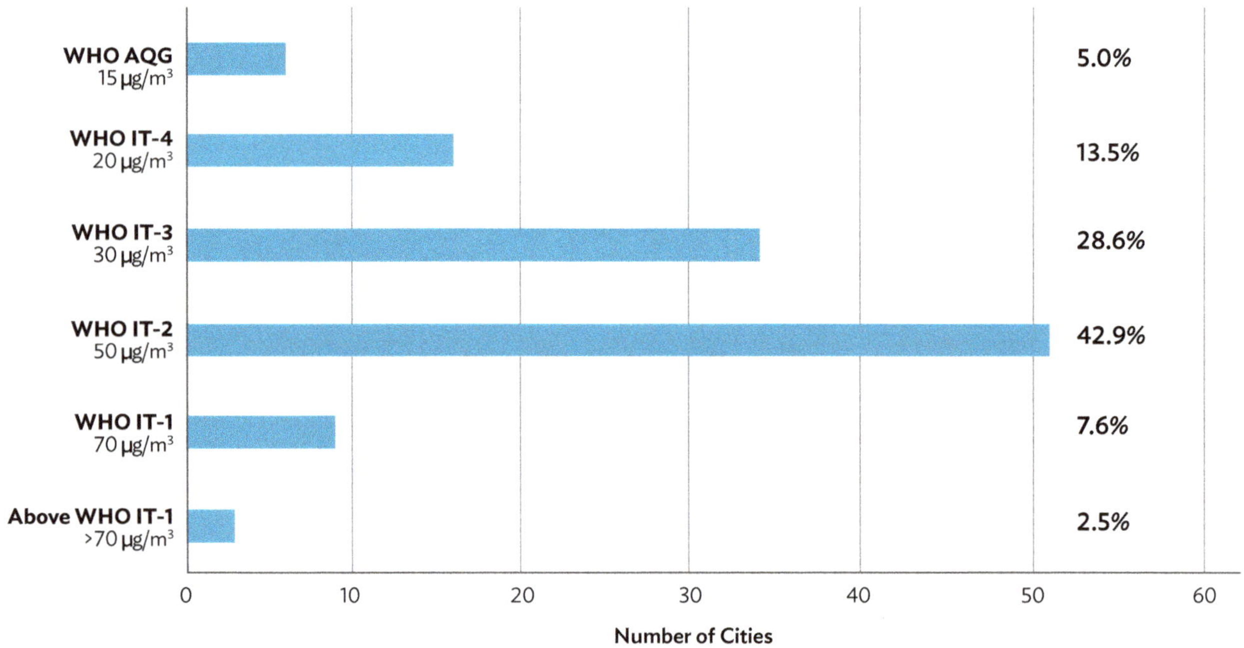

AQG = Air Quality Guidelines, ASEAN = Association of Southeast Asian Nations, IT = interim target, PM = particulate matter, WHO = World Health Organization.

Source: Clean Air Asia database on the 2024 annual average PM_{10} concentrations in Asian cities.

Figure 6: Distribution of ASEAN Cities by 2024 PM_{10} Levels Across WHO AQG and Interim Targets, by Country

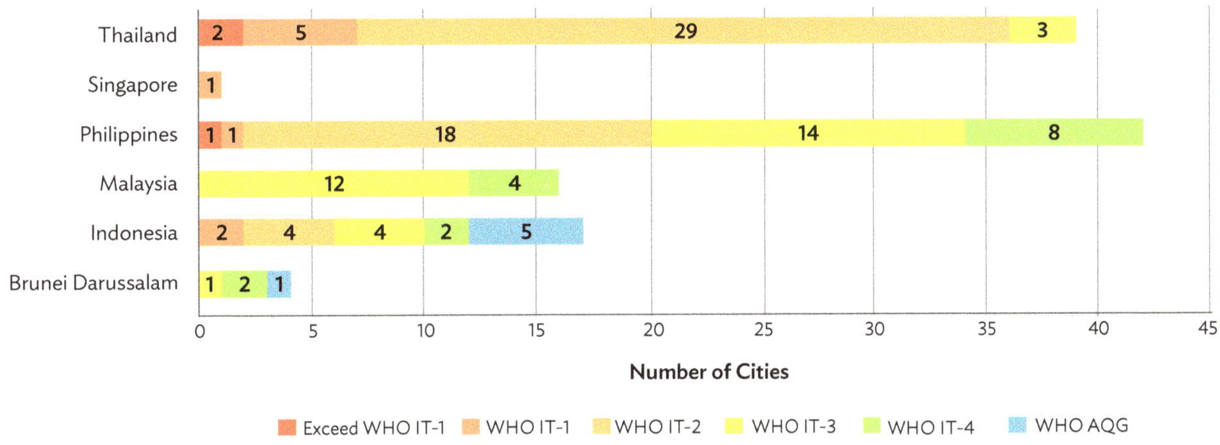

AQG = Air Quality Guidelines, ASEAN = Association of Southeast Asian Nations, IT = interim target, PM = particulate matter, WHO = World Health Organization.

Source: Clean Air Asia database on the 2024 annual average PM_{10} concentrations in Asian cities.

Nitrogen Dioxide Levels and Monitoring Coverage in 2024

Nitrogen dioxide (NO_2) monitoring coverage across ASEAN remains limited relative to $PM_{2.5}$. In 2024, NO_2 data were available in only 41 cities across half of the ASEAN member states, constraining a comprehensive regional assessment. Figure 7 shows that only nine cities recorded annual average concentrations that met WHO AQG of 10 $\mu g/m^3$, indicating that 78% of cities exceeded the recommended level. Most of the remaining cities fell within WHO IT-3 (20 $\mu g/m^3$) level, the next most stringent threshold. In Figure 8, disaggregated results show that only cities in Brunei Darussalam, Malaysia, and the Philippines met WHO AQG.

The limited availability of NO_2 data highlights a significant monitoring gap in the region. As a key marker of combustion-related pollution, mainly from vehicle emissions and industrial processes, NO_2 provides important insight into urban air quality conditions.

Expanding monitoring coverage, particularly in densely populated and traffic-congested urban centers, would strengthen the evidence base for emission control strategies, improve trend analysis, enhance evaluation of policy effectiveness, and protect public health.

Figure 7: Distribution of ASEAN Cities by 2024 NO_2 Average Across WHO AQG and Interim Targets

AQG = Air Quality Guidelines, ASEAN = Association of Southeast Asian Nations, IT = interim target, NO_2 = nitrogen dioxide, WHO = World Health Organization.
Source: Clean Air Asia database on the 2024 annual average NO_2 concentrations in Asian cities.

Figure 8: Distribution of ASEAN Cities by 2024 NO_2 Average Across WHO AQG and Interim Targets, by Country

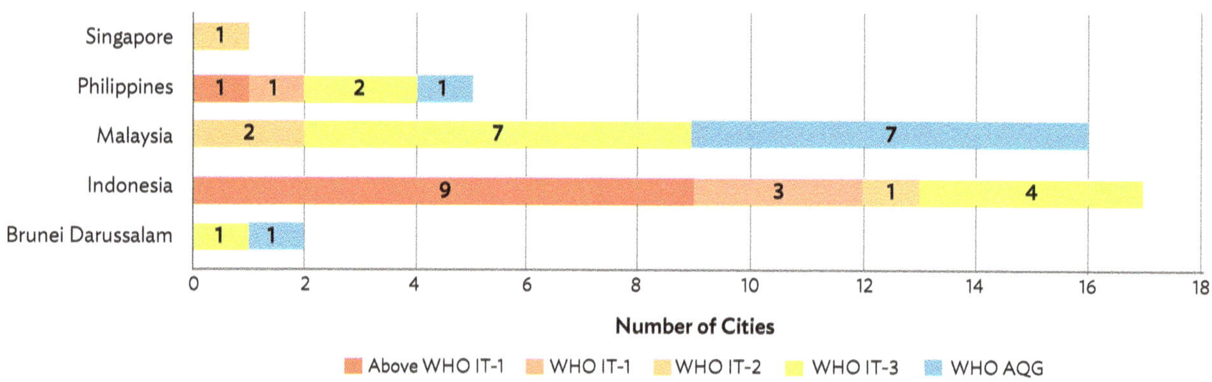

AQG = Air Quality Guidelines, ASEAN = Association of Southeast Asian Nations, IT = interim target, NO_2 = nitrogen dioxide, WHO = World Health Organization.
Source: Clean Air Asia database on the 2024 annual average NO_2 concentrations in Asian cities.

Air Quality in ASEAN's Largest Cities

As the region's largest urban centers, ASEAN's most populous cities play a critical role in shaping regional air quality outcomes. These cities experience the highest levels of traffic, energy use, and industrial activity, and are often where public health impacts are most concentrated. Monitoring and addressing air quality in these centers will provide important insight into broader challenges and opportunities for sustainable urban development across ASEAN.

A man charges his Vinfast electric car at a charging station in a new urban area in Ha Noi, Viet Nam (photo by ADB).

The most populous city in each country that is included in this report provides a representative view of urban air quality conditions in the region (2024 population estimates): [2]

- Metro Manila, Philippines (14.9 million)
- Jakarta, Indonesia (11.4 million)
- Bangkok, Thailand (11.2 million)
- Ho Chi Minh City, Viet Nam (9.6 million)
- Kuala Lumpur, Malaysia (8.8 million)
- Singapore, Singapore[3] (6.1 million)
- Yangon, Myanmar (5.7 million)
- Phnom Penh, Cambodia (2.3 million)
- Vientiane, Lao People's Democratic Republic (0.7 million)
- Bandar Seri Begawan, Brunei Darussalam (0.1 million)

Figure 9 shows that none of the cities with available 2024 $PM_{2.5}$ data met the WHO Air Quality Guideline (AQG) of 5 $\mu g/m^3$.[4] Bandar Seri Begawan was the only city to meet WHO IT-4 (10 $\mu g/m^3$), while Singapore met WHO IT-3 (15 $\mu g/m^3$). Phnom Penh, Kuala Lumpur, Ho Chi Minh City, and Metro Manila met Interim Target-2 (IT-2) of 25 $\mu g/m^3$, while the remaining cities exceeded this threshold. These findings highlight persistent challenges across major urban centers, particularly in Vientiane, Jakarta, Bangkok, and Yangon, where concentrations remain well above the more stringent WHO interim targets and WHO AQG.

2 United Nations. 2025. Population, Surface Area and Density. *Statistical Yearbook No. 68*.

3 Singapore is a city-state and given its small size and urbanized nature, the state is treated as comparable to other major cities in the region for the purposes of this analysis.

4 2024 $PM_{2.5}$ data is used in the analysis as they represent the most recent available dataset.

Figure 9: Annual $PM_{2.5}$ Levels in Most Populated ASEAN Cities Relative to WHO AQG and Interim Targets

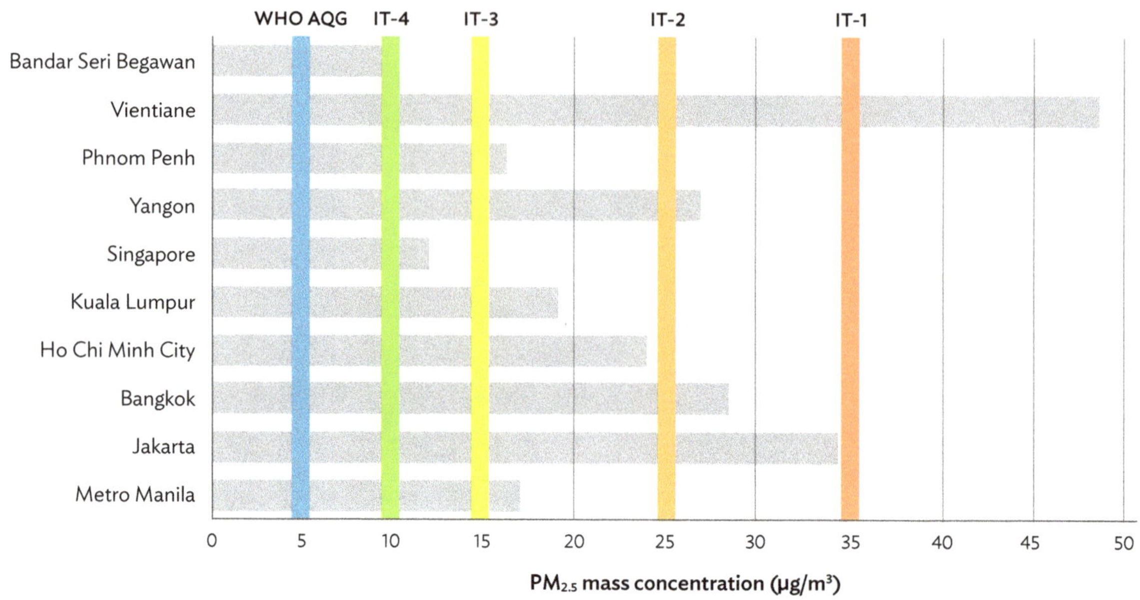

AQG = Air Quality Guidelines, ASEAN = Association of Southeast Asian Nations, IT = interim target, PM = particulate matter, WHO = World Health Organization.

Sources: Data for all cities were obtained from the Clean Air Asia database, with most values updated in 2024, except for Metro Manila (2022), Ho Chi Minh City (2022), and Yangon (2023).

Achieving air quality aligned with WHO AQG will require leadership and sustained coordination between national and municipal authorities. Densely populated urban areas are major emission sources and offer significant opportunities for policy impact. City governments play a central role not only in implementing national regulations, but also in designing locally tailored interventions. Actions such as traffic management, public transport improvements, waste system upgrades, and targeted emission controls can materially influence urban air quality outcomes. Empowering cities to lead in air quality management—supported by adequate institutional and financial resources—will be critical to building healthier, more resilient, and sustainable urban environments across the region.

2 Air Quality Monitoring Capacity and Gaps

Understanding air quality in the region requires examining pollution levels and assessing how these levels are measured. The reliability, coverage, and accessibility of monitoring systems directly influence how air quality trends are interpreted and how policy responses are designed. This section reviews the current state of air quality monitoring (AQMt) across the Association of Southeast Asian Nations (ASEAN), including monitoring approaches, network coverage, and implications for air quality management.

Effective air quality management depends on robust monitoring systems. An AQMt network generates the data necessary to inform evidence-based decision-making and evaluate progress toward national standards and international guidelines. According to the Clean Air Asia Guidance Framework and City Solutions Toolkit,[5] monitoring networks serve the following core objectives:

- **Timely public reporting.** Provides real-time or short-term air pollution data to support air quality index (AQI) reporting, forecasting, and activation of response protocols during pollution episodes.
- **Compliance and/or planning.** Compares measured concentrations with national standards, guidelines, and other policy targets to support regulatory enforcement and policy development.
- **Impact assessment.** Evaluates the effects of air pollution on human health, ecosystems, agriculture, and the environment, and assesses the effectiveness of AQMt policies.
- **Long-term monitoring.** Establishes background pollution concentration levels and long-term trend analysis to inform future planning and strategy development.

The scope and configuration of AQMt networks vary according to national priorities, monitoring objectives, institutional capacity, and available resources. Monitoring systems may serve multiple objectives simultaneously. Given the financial and technical requirements involved, resource availability often influences the scale, density, and technological sophistication of national monitoring networks.

All ASEAN member states operate some form of AQMt. Brunei Darussalam primarily relies on fixed monitoring stations, while most other countries employ hybrid systems combining fixed (reference and non-reference instruments) and mobile platforms, including both continuous and filter-based monitoring methods.

Population-Based Monitoring Adequacy

Monitoring network size should be assessed relative to the population served. Adequate population-based coverage is essential to ensure representative air quality data and to link measured concentrations to exposure across entire populations, including vulnerable groups. Population size and density therefore play a central role in determining the appropriate number and distribution of monitoring stations.

To evaluate monitoring adequacy, this report analyzed the population of 10 ASEAN countries, the current number of government-owned monitoring stations, and the recommended number of monitoring stations based on the European Union Directive 2008/50/EC as illustrated in Table 1.[6] The directive provides internationally recognized guidance on minimum monitoring density relative to population and concentration thresholds and is used here as a benchmark for comparison.

5 Clean Air Asia. 2020. IBAQ City Solutions Tool Kit, Guidance Area 1.

6 European Union (EU). 2008. Directive 2008/50/EC of the European Parliament and of the Council on Ambient Air Quality and Cleaner Air for Europe. *Official Journal of the European Union.*

Table 1: Prescribed Sampling Points from the European Union Directive 2008/50/EC

Population Aggregation or Zone ('000)	If the maximum concentrations exceed the upper assessment threshold		If the maximum concentrations are between the upper and lower assessment thresholds	
	Pollutants Except PM	PM (PM_{10} + $PM_{2.5}$)	Pollutants Except PM	PM (PM_{10} + $PM_{2.5}$)
0–249	1	2	1	1
250–499	2	3	1	2
500–749	2	3	1	2
750–999	3	4	1	2
1,000–1,499	4	6	2	3
1,500–1,999	5	7	2	3
2,000–2,749	6	8	3	4
2,750–3,749	7	10	3	4
3,750–4,749	8	11	3	6
4,750–5,999	9	13	4	6
≥ 6,000	10	15	4	7

PM = particulate matter.
Source: The European Union Directive 2008/50/EC.

Figure 10 shows that Brunei Darussalam, Cambodia, and Singapore have monitoring networks broadly aligned with population-based benchmarks. Table 2 provides the corresponding values, showing that other countries operate fewer government-owned stations than suggested by the benchmark guidance referenced above. While government-owned stations form the backbone of national monitoring systems, private sector and civil society initiatives have also contributed to expanding coverage. Greater coordination among these groups may help strengthen overall monitoring capacity and improve data representativeness.

Figure 10: Monitoring Sites in ASEAN Countries Relative to Population

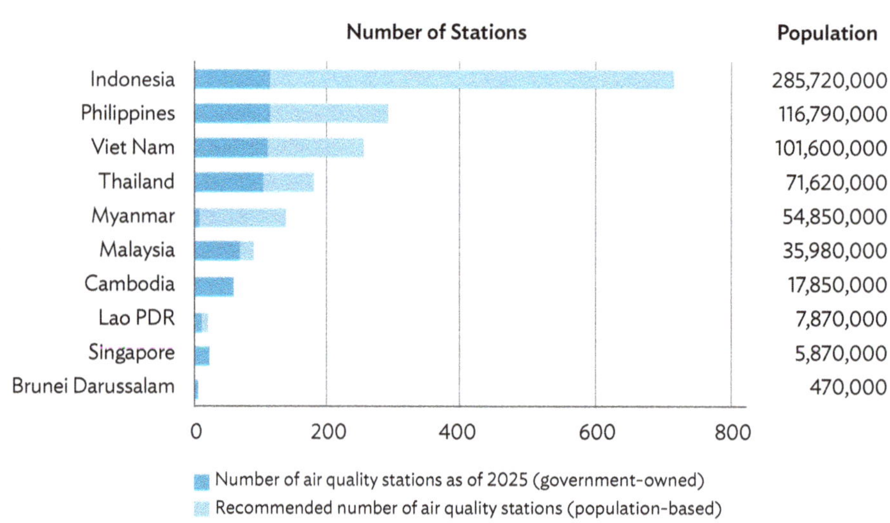

ASEAN = Association of Southeast Asian Nations, Lao PDR = Lao People's Democratic Republic.

Notes: For countries with a population greater than 6 million, the ratio of 1 monitoring station per 400,000 people was used. This ratio is derived from the last tier of recommendations for particulate matter sampling under conditions where maximum concentrations exceed the upper assessment threshold (Table 1). Under this condition, the European Union Directive 2008/50/EC prescribes 15 monitoring stations for populations of 6 million or more. This ratio was used as the basis for estimating the recommended number of monitoring stations for a given population.

Source: Population values are based on data from the United Nations Population Division.

Table 2: Government-Owned vs. Recommended Monitoring Sites in ASEAN Countries

Country	Government-Owned AQM Stations	Recommended AQM Stations (Population-Based)
Brunei Darussalam	6	3
Cambodia	59	45
Indonesia	113	714
Lao PDR	10	20
Malaysia	68	90
Myanmar	7	137
Philippines	113	292
Singapore	22	13
Thailand	104	179
Viet Nam	111	254

AQM = Air Quality Monitoring, ASEAN = Association of Southeast Asian Nations, Lao PDR = Lao People's Democratic Republic.
Source: Clean Air Asia.

Determinants of Monitoring Network Size

Population appears to be a stronger determinant of monitoring network size than national income level. Indonesia, the Philippines, and Viet Nam—classified as upper- and lower-middle-income economies—operate more government-owned monitoring stations than some higher-income countries. This suggests that population scale and associated exposure pressures may drive investment in monitoring infrastructure, even in resource-constrained settings. People in low- and middle-income countries bear the highest health burden from outdoor air pollution, reinforcing the urgency of strengthening monitoring systems in these contexts.[7]

Geographic and urban characteristics further shape monitoring strategies. Singapore, for example, operates fewer stations despite having the highest population density in the region. Its relatively small land area and centralized urban layout allow broader coverage with fewer strategically placed monitors. However, dense urban environments often involve complex patterns of emissions, traffic flows, land use activities, and micro-scale atmospheric conditions, which can result in significant spatial variability of pollution concentrations even within small areas. Monitoring network design must therefore consider not only population size, but also urban form and emission patterns. These relationships are illustrated in Figure 11.

Figure 11: Monitoring Sites in ASEAN Countries Relative to Population, Population Density, and Income

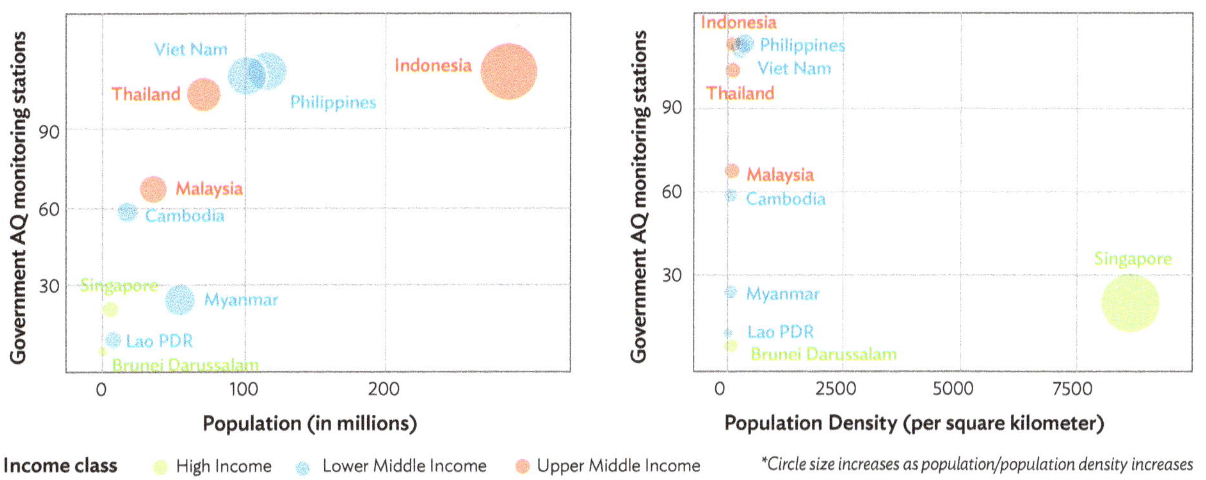

AQ = air quality, ASEAN = Association of Southeast Asian Nations, Lao PDR = Lao People's Democratic Republic.
Note: The graphs include information from the 10 ASEAN countries.
Source: Country income data were from the World Bank's World Development Indicator.

[7] WHO. 2022. Ambient (outdoor) Air Quality and Health.

Another factor associated with monitoring expansion is air pollution mortality, as shown in Figure 12. Cambodia, the Lao PDR, and Myanmar report some of the highest air pollution-related mortality per 100,000 population in the region. In such contexts, expanded monitoring coverage can improve understanding of exposure patterns and support more targeted air quality management responses.

Figure 12: Monitoring Sites in ASEAN Countries Relative to Air Pollution Mortality, Population, and Income

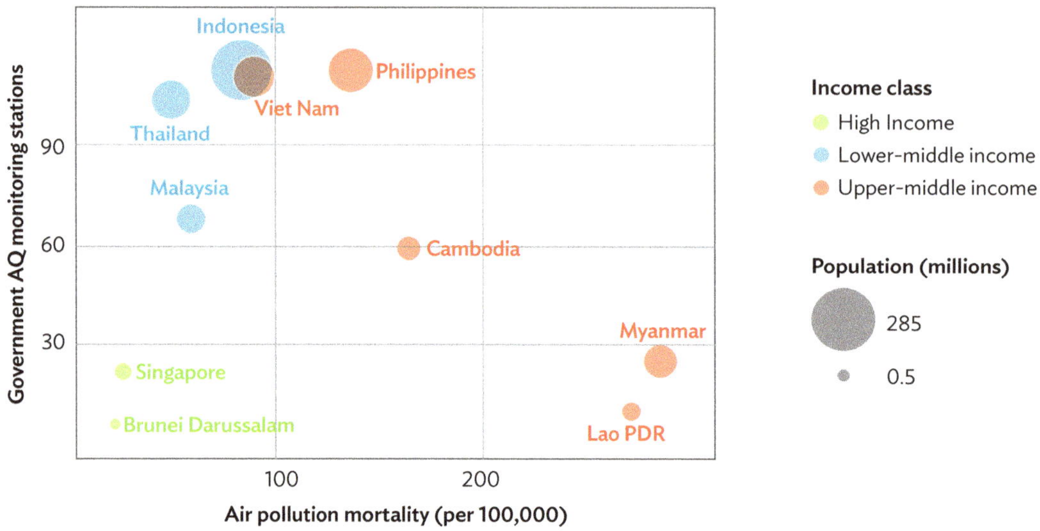

AQ = air quality, Lao PDR = Lao People's Democratic Republic.
Sources: Data were from the Clean Air Asia database based on publicly accessible government websites, reports, and documents. Mortality data were based on the Global Burden of Disease 2023 released by the Health Effects Institute in 2025. Population data were from the UN Population Division.

In countries with limited institutional or financial capacity, hybrid monitoring approaches can provide a practical pathway to expand air quality monitoring coverage.[8] These systems combine reference-grade instruments with calibrated low-cost sensors and satellite-based measurements while maintaining data quality standards.[9] Such approaches can improve spatial coverage at lower cost, particularly in lower-income settings where establishing extensive reference-grade monitoring networks may not be immediately feasible.

Monitoring Coverage by Pollutant

Monitoring across ASEAN remains concentrated on particulate matter (PM), particularly $PM_{2.5}$ and PM_{10}. Coverage of other pollutants, such as ground-level ozone (O_3), nitrogen dioxide (NO_2), sulfur dioxide (SO_2), black carbon (BC), elemental carbon (EC), and ultrafine particles ($PM_{0.1}$), is comparatively limited. Each pollutant has distinct health and environmental impacts, and understanding their complex interactions is critical for developing integrated air quality management strategies.

The 2021 World Health Organization (WHO) Air Quality Guidelines (AQG) emphasize the importance of monitoring a broader range of pollutants, including BC, EC, and ultrafine particles. While monitoring of these pollutants has advanced in parts of Europe and other regions, coverage in ASEAN remains uneven. Broader pollutant monitoring would enhance understanding of emission sources, exposure pathways, and the effectiveness of control strategies.

8 Hybrid approach or system in AQM refers to a combination of conventional (e.g., reference instruments) and unconventional (e.g., low-cost sensors and satellite-based techniques) methods.

9 Clean Air Asia. 2020. *City Solutions Toolkit: Technology Options for Air Quality Monitoring: Conventional vs. Next-generation*.

Monitoring BC is also relevant for climate policy, as it is a short-lived climate pollutant that contributes to both adverse health effects and global warming.

In 2024, $PM_{2.5}$ was the most widely monitored pollutant across ASEAN, followed by PM_{10}. Indonesia reported the most comprehensive pollutant coverage including all criteria air pollutants from multiple cities, followed by Malaysia. Monitoring of NO_2, carbon monoxide (CO), and O_3 remains limited in several countries, while SO_2 coverage is more constrained. Although PM serves as a reliable indicator of air quality and associated health risks, incorporating additional pollutants provides a more complete assessment of air pollution impacts on health, ecosystems, and the climate. These differences in pollutant coverage are illustrated in Figure 13.

Figure 13: Air Pollutants Monitored and Reported in ASEAN Countries, 2024

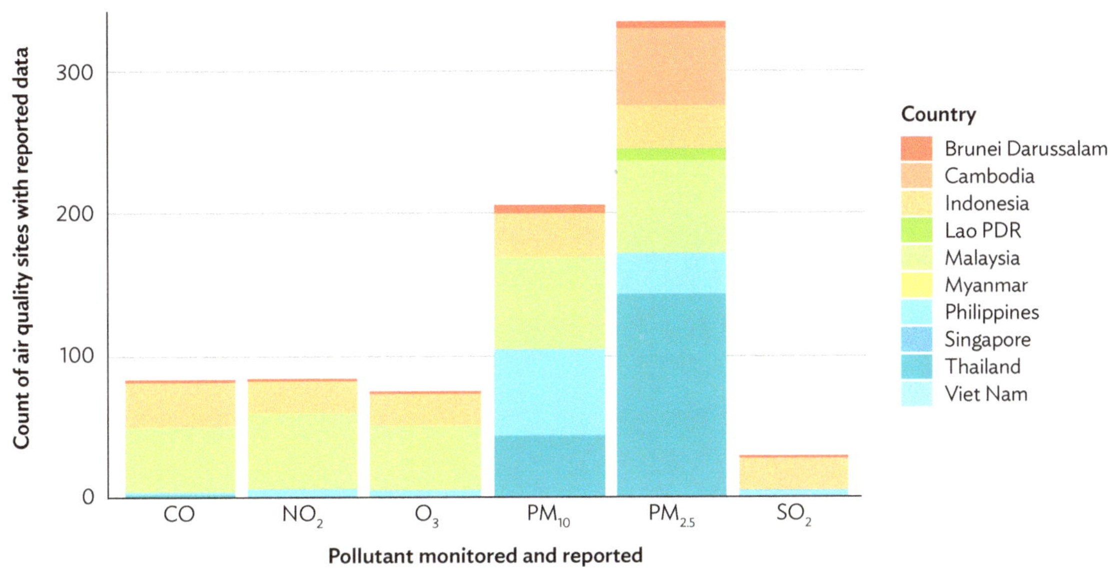

CO = carbon monoxide, Lao PDR = Lao People's Democratic Republic, NO_2 = nitrogen dioxide, O_3 = ground-level ozone, PM_{10} = particulate matter$_{10}$, $PM_{2.5}$ = particulate matter, SO_2 = sulfur dioxide.

Source: Clean Air Asia.

Data Accessibility

The effectiveness of monitoring systems depends not only on the number of stations, but also on the accessibility of the data generated. Public access to real-time and historical air quality data enhances transparency, supports research and policy analysis, and strengthens public trust. Accessibility also enables stakeholders, including community leaders, researchers, civil society, and local authorities, to engage more effectively in air quality management.

While the number of AQMt stations serves as an indicator of monitoring capacity, a gap often exists between the number of installed stations and publicly available datasets. In some cases, official reports do not specify whether stations are operational or inactive during a given period. For consistency, this report assumes that all reported stations are producing data unless otherwise noted. Greater transparency regarding station status and data capture rates would provide clearer insight into maintenance needs and investment requirements.

To assess accessibility, this report compared the number of AQMt stations with publicly accessible $PM_{2.5}$ data to the total number of government-owned stations in 10 ASEAN countries.[10] $PM_{2.5}$ was selected given its well-documented health impacts and widespread monitoring across the region.

[10] The report refers to "available" air quality data if the information exists, and "accessible" air quality data if the information can be obtained easily by the public. Please refer to Appendix 1: Methodology for more information.

Indonesia (113), the Philippines (113), Viet Nam (111), and Thailand (104) operate the largest number of government-managed monitoring stations (manual and automatic). Most countries publish real-time air quality information through national indices such as AQI or Pollutant Standards Index (PSI), though only three provide real-time concentration values (Cambodia, Singapore, and Thailand). Thailand stands out for providing detailed hourly and historical data through its public Air4Thai platform (https://air4thai.pcd.go.th). Across the region, however, access to comprehensive historical data remains limited in several countries. For 2024, publicly accessible city-level $PM_{2.5}$ data are available only in a small number of countries, notably Thailand and Singapore, while several others provide historical datasets only up to 2022 (Figure 14).

Delays or gaps in public data availability constrain the ability to track trends, assess policy effectiveness, and conduct independent analysis. Improving the timeliness and completeness of data dissemination would strengthen regional transparency and support more evidence-based air quality management.

Figure 14: Monitoring Sites in ASEAN Countries with Accessible $PM_{2.5}$ Data, 2022

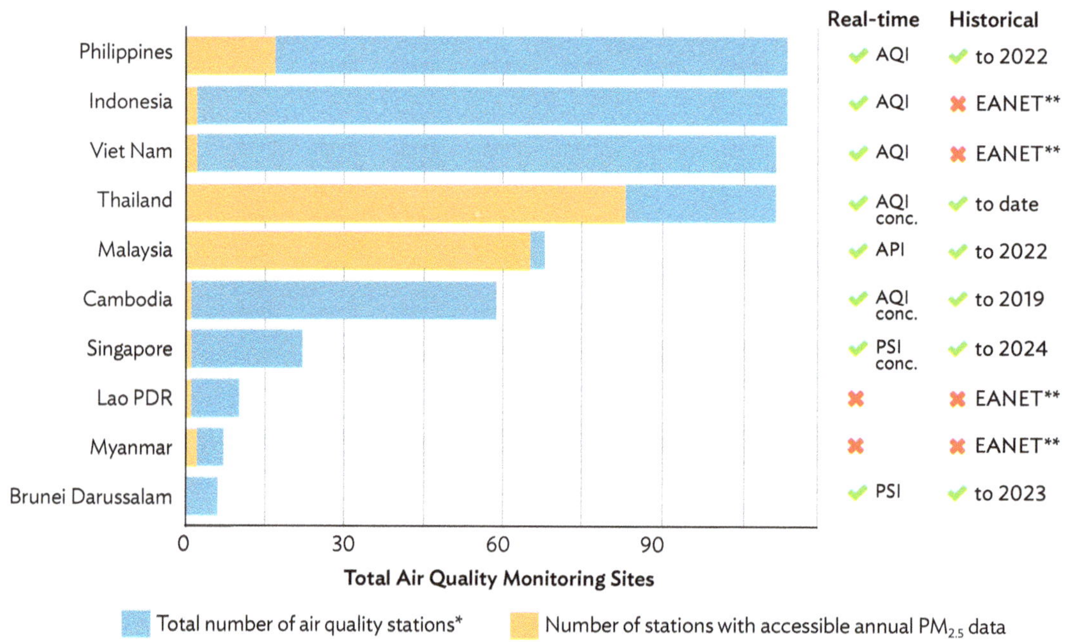

API = air pollution index, AQI = air quality index, EANET = Acid Deposition Monitoring Network in East Asia, Lao PDR = Lao People's Democratic Republic, PM = particulate matter, PSI = pollutant standards index.

Sources:

* Clean Air Asia database as of June 2025; but the year of the most recent available data varies per country (e.g., 2022 is the year with the most data). For 2024, publicly accessible city-level data are available only for a limited number of countries (Singapore and Thailand).

** Historical data is available from the Acid Deposition Monitoring Network in East Asia (EANET), which is not an official government channel.

Taken together, the analysis indicates that while all ASEAN countries have established some form of air quality monitoring, significant gaps remain in coverage, pollutant scope, and data accessibility. Monitoring capacity varies widely relative to population size, exposure risk, income level, and mortality burden. Expanding and diversifying monitoring networks, while ensuring transparency and data quality, will be critical to strengthening evidence-based air quality management across the region. Robust and accessible monitoring systems not only support policy evaluation and regulatory enforcement but also provide the foundation for informed public engagement and long-term environmental planning.

3 Sectoral Emission Sources

Sustaining air quality improvements across the Association of Southeast Asian Nations (ASEAN) requires a clear understanding of the main drivers of pollution. Countries are therefore encouraged to strengthen the development of comprehensive national and subnational emission inventories and source apportionment studies. Effective mitigation depends on identifying where emissions originate, quantifying their magnitude, and determining which sectors pose the greatest risks to human health and the environment.

To support a more coordinated regional approach to air quality management, ASEAN countries are encouraged to adopt harmonized protocols for developing emission inventories. A regional emissions profile remains challenging due to methodological differences, varying data quality, and inconsistent reporting practices across countries. Standardized templates for activity data, a phased transition to higher-tier methodologies, and the development of region-specific emission factors could improve the consistency and comparability of emission inventories. Sharing best practices in conducting source apportionment studies would further strengthen regional cooperation. A more aligned approach would enable coordinated planning and more effective action across borders.

In ASEAN, only a few countries publish their national air pollutant inventories through official websites or reports, including Brunei Darussalam,[11] the Philippines,[12] and Singapore.[13] Other organizations and networks provide supplementary information on emission sources for several countries, including Cambodia,[14] Indonesia,[15] the Lao PDR,[16] Malaysia,[17] Thailand,[18] Viet Nam,[19] as well as the region as a whole.[20] Academic institutions and nongovernment organizations also play a key role in the development of emission inventories and source apportionment studies at the city level, with findings shared through scientific publications and online platforms. All ASEAN member states have developed greenhouse gas (GHG) inventories under their reporting obligations to the United Nations Framework Convention on Climate Change.

A global study,[21] using satellite-derived particulate matter ($PM_{2.5}$) concentrations and emissions modelling, indicates that residential sources are the largest contributors to ambient $PM_{2.5}$ concentration globally, accounting for more than 25%, primarily from solid fuel use. Emissions from the energy and industry sectors follow, contributing some 22% (including both coal and non-coal-based activities). The transport sector contributes around 10%, while the agriculture and waste sectors each contribute around 5%.

---

11 Government of Brunei Darussalam, Ministry of Development. n.d. Pollution in Brunei Darussalam. (Only describes main sources of pollution).

12 Government of the Philippines, Department of Environment and Natural Resources, Environmental Management Bureau. 2021. Emissions Inventory 2021. (dashboard of all criteria air pollutants).

13 Government of Singapore, National Environment Agency. 2023. Air Quality in Singapore. (only describes SO_2 EI).

14 Climate & Clean Air Coalition. n.d. Clean Air Plan of Cambodia.

15 Systemiq. 2025. *Better Air, Better Indonesia: The Economic and Political Case for Urgent and Coordinated Action for Jakarta's Clean Air*.

16 O'Neill, C. 2024. Evidence for Clean Air Action: Quantifying Emission Sources in Lao PDR. Stockholm Environment Institute.

17 Shafie, S. H. M. et al. 2022. Influence of Urban Air Pollution on the Population in the Klang Valley, Malaysia: A Spatial Approach. *Ecological Processes*. 11(1).

18 ChooChuay, C. et al. 2020. Impacts of $PM_{2.5}$ Sources on Variations in Particulate Chemical Compounds in Ambient Air of Bangkok, Thailand. *Atmospheric Pollution Research*. 11(9). pp. 1657–1667.

19 Ho, B. Q. et al. 2021. Photochemical Modeling of $PM_{2.5}$ and Design Measures for $PM_{2.5}$ Reduction: A Case of Ho Chi Minh City, Viet Nam. *IOP Conference Series: Earth and Environmental Science*. 652(1). 012025.

20 Reddington, C. L. et al. 2019. Exploring the Impacts of Anthropogenic Emission Sectors on $PM_{2.5}$ and Human Health in South and East Asia. *Atmospheric Chemistry and Physics*. 19(18). pp. 11887–11910.

21 McDuffie E. et al. 2021. Fine Particulate Matter and Global Health: Fuel and Sector Contributions to Ambient $PM_{2.5}$ and its Disease Burden Across Multiple Scales. *Nature Communications*.

Figure 15: Source Sectors of Ambient $PM_{2.5}$ in Asia

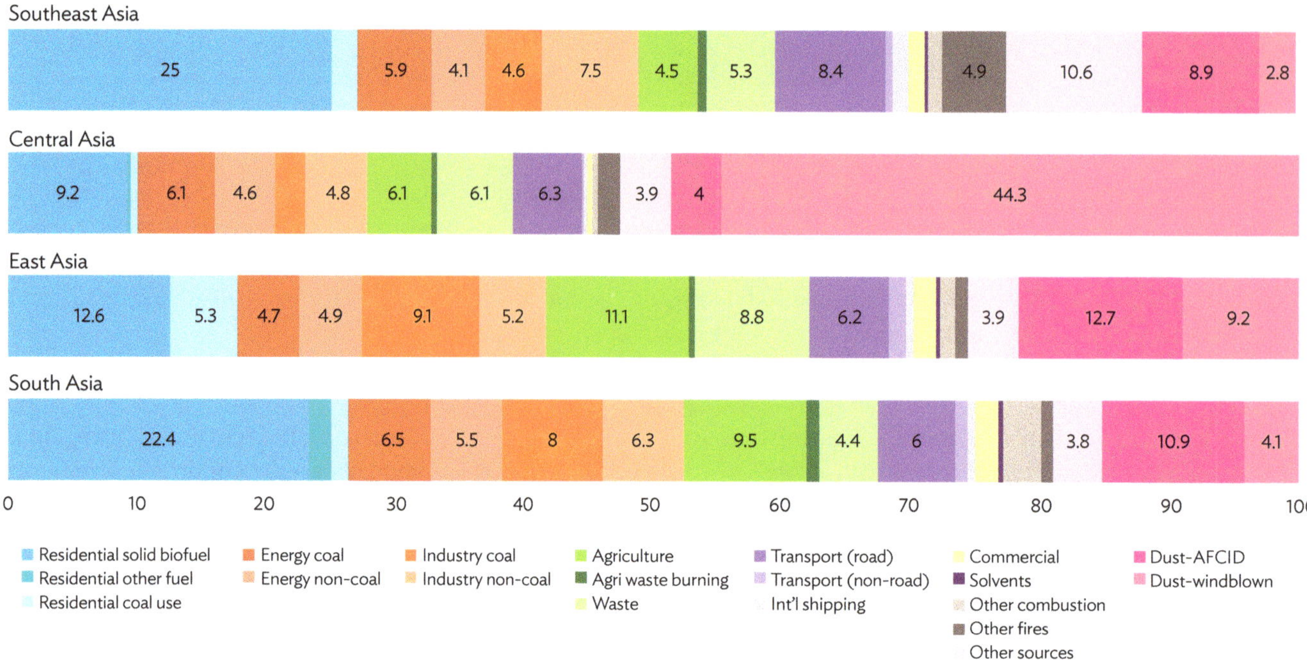

AFCID = anthropogenic fugitive, combustion, and industrial dust; PM = particulate matter.

Source: McDuffie, E.E. et al. 2021. Source Sector and Fuel Contributions to Ambient $PM_{2.5}$ and Attributable Mortality Across Multiple Spatial Scales. *Nature Communications*. 12: 3594. https://doi.org/10.1038/s41467-021-23853-y.

While results of the global modelling shown in Figure 15 point to residential solid biofuel as the main source of $PM_{2.5}$ in Southeast Asia, information from publicly accessible national emissions inventories and source apportionment studies suggests variations across countries, as illustrated in Figure 16. In four ASEAN member countries, transport is identified as the leading source of $PM_{2.5}$ emissions. In three other countries, industry is reported as the primary contributor. Biomass burning and area sources are dominant in several remaining countries. However, the compiled information does not consistently include emission magnitudes due to data inaccessibility. National inventories provide useful country-level insights for sectoral prioritization, but more granular subnational assessments are needed to guide subnational or city-level interventions. For regional interpretation, greater methodological harmonization is essential.

Emission profiles for other pollutants should also be considered when designing mitigation strategies. Some sectors may not be dominant contributors to $PM_{2.5}$ but can significantly affect other pollutants. Agriculture and waste sectors, for example, are major sources of ammonia and methane, respectively. Even within a single sector, emissions vary depending on fuel types, technologies, and equipment. Emission inventories should therefore be pollutant-specific and disaggregated by detailed source categories. Incorporating geospatial approaches can further improve understanding by accounting for the location and concentration of emission sources within defined boundaries.

A clearer understanding of sectoral emission profiles enables governments to prioritize high-impact interventions and allocate resources more effectively. Variations across ASEAN countries underscore that mitigation strategies must reflect national and local conditions rather than follow a uniform model. Strengthening emissions analysis will therefore be central to translating regional air quality goals into targeted and measurable reductions.

Figure 16: Major Source Sectors of $PM_{2.5}$ in ASEAN Countries Based on National Emission Inventories and Local Studies

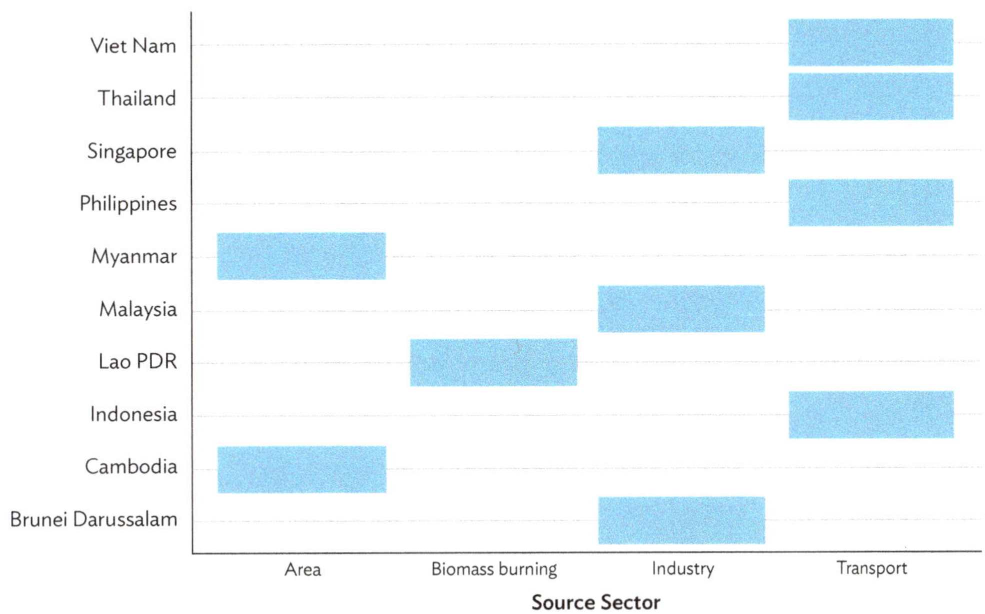

ASEAN = Association of Southeast Asian Nations, Lao PDR = Lao People's Democratic Republic, PM = particulate matter.
Source: Clean Air Asia.

4 Air Quality Governance and Policy Frameworks

The knowledge base on air pollution levels, emission sources, and impacts continues to expand, enabling countries in the region to make informed decisions to improve air quality. This creates an opportunity to translate growing technical evidence into stronger governance frameworks and more effective policy implementation.

This section highlights key legal frameworks and strategies implemented across member states of the Association of Southeast Asian Nations (ASEAN). It outlines national and subnational plans and actions aimed at reducing emissions, strengthening standards, integrating air quality and public health management, and improving communication with the public.

Laws, Plans, and Strategies

ASEAN countries are encouraged to periodically review and amend national-level legal frameworks to ensure continued relevance, considering emerging scientific evidence, technological advancements, and evolving socioeconomic conditions. Legal frameworks serve as the foundation for developing, implementing, and sustaining air quality management systems. They underpin the establishment of air quality standards and guidelines, define institutional mandates for monitoring and enforcement, and provide the legal basis for regulating emissions and penalizing violators. These frameworks also delineate roles and responsibilities across levels of government, facilitating coordination and communication among agencies.

Of the 10 countries included in this report, four have a primary law specifically addressing air quality, while the remaining six rely on broader environmental protection laws to govern air quality management. At a minimum, these laws contain sectoral provisions to address air pollution and improve air quality, while some also integrate climate change considerations. The earliest environmental law in the region was Malaysia's Environmental Quality Act *(1974)*, followed by environmental protection laws in the Lao People's Democratic Republic (Lao PDR) and Singapore in 1999.[22] The first air quality-specific laws were introduced in Indonesia and the Philippines in 1999. Malaysia's law was amended in 2024, while Singapore revised its law in 2020 and the Lao PDR in 2013.

To complement legislation, countries have adopted a range of policies, action plans, and sector-specific programs aimed at improving air quality, as shown in Table 3. These instruments operationalize legal mandates and provide greater implementation detail. National action plans typically outline strategies for reducing emissions across key sectors such as transport, energy, and industry. In some cases, they include quantitative reduction targets for specific air pollutants and corresponding timelines. Several countries have also introduced sectoral policies, including stricter vehicle emission standards, improved fuel quality requirements, and updated industrial emission regulations. Some policies further integrate climate objectives, recognizing the co-benefits of reducing air pollutants and greenhouse gas (GHG) emissions. Singapore has enacted a dedicated law on transboundary haze pollution, outlining protocols for emergency response and enforcement during haze episodes. This demonstrates how countries can tailor governance frameworks to address specific national or regional challenges. Whether cross-cutting or sector-specific, these policy instruments are critical for translating high-level legislation into actionable and measurable progress on air quality.

---

22 Singapore had a Clean Air Act in 1971 but was repealed after the Parliament passed the Environmental Protection and Management Act in 1999.

Table 3: National Air Quality Policies and Plans in ASEAN Countries

Country	National Policy and/or Action Plan	Year	Air Quality Targets	Sectoral Actions	AQ-Climate Integration	Haze/AP Episode Protocol
Brunei Darussalam	**Environmental Protection and Management Order 2016**	2016	✖	✔	✖	✖
Cambodia	**Sub-Decree No. 42 on Air Pollution Control and Noise Disturbance**	2000	✖	✔	✖	✖
	Clean Air Plan of Cambodia	2021	✔	✔	✔	✖
	Air Quality Improvement Program (2023–2025)	2023	✖	✔	✔	✖
	Circular No. 1 on Measures to Prevent and Reduce Ambient Air Pollution (Revised $PM_{2.5}$ and PM_{10} Standards)	2020	✔	✔	✔	✖
	Improved VES: Euro IV for Second-Hand Vehicles by 2022, Euro V by 2027	2022	✔	✔	✔	✖
	Circular Strategy on Environment	2023	✖	✔	✔	✖
	Code on Environmental and Natural Resources	2023	✖	✔	✔	✖
Indonesia	**Government Regulation No. 41/1999: Control of Air Pollution**	1999	✖	✔	✖	✖
	Governor Regulation No. 22 of 2021 on Environmental Protection, Organization, and Management	2021	✖	✖	✔	✖
	MOEF Regulation P. 20: Euro IV Standard for Diesel Vehicles	2022	✖	✔	✖	✖
Lao PDR	**Environmental Protection Law 1999** (revised 2013)	2013	✖	✔	✔	✖
	Health National Adaptation Plan	2024	✖	✔	✔	✖
Malaysia	**Environmental Quality Act 1974** (amended 2024)	2024	✖	✔	✖	✖
	New Malaysia Ambient Air Quality Standard	2020	✖	✖	✖	✖
	Euro V Diesel Implementation	2021	✖	✔	✖	✖
	Environmental Quality (Clean Air Regulations)	2014	✖	✔	✖	✖
Myanmar	**Environmental Conservation Law 2012**	2012	✖	✔	✔	✖
Philippines	**Republic Act No. 8749: Philippine Clean Air Act 1999**	1999	✖	✔	✔	✖
	DENR Administrative Order No. 2020-14	2020	✖	✖	✖	✖
	Integrated Air Quality Improvement Framework–Air Quality Control Action Plan for 2025–2035 (draft)	-	✖	✔	✔	✖
Singapore	**Environmental Protection and Management Act 1999** (revised 2020)	2020	✖	✔	✖	✖
	Transboundary Haze Pollution Act 2014 (revised 2020)	2020	✖	✔	✖	✔
	Enhanced and Commercial VES	2021	✖	✔	✔	✖
	Carbon Pricing Act Amendment	2022	✖	✔	✔	✖
	Singapore Green Plan 2030	2021	✔	✔	✔	✖
Thailand	**Clean Air Act** (draft)	-	-	✔	-	-
	Revised Ambient Air Quality Standard for $PM_{2.5}$	2023	✖	✖	✖	✖
	Euro V for Diesel by 2024, Euro VI for Light Duty Vehicles by 2025	2023	✖	✔	✖	✖
	Standards B.E. 2566 for controlling air emissions from powerplants	2023	✖	✔	✖	✖

continued on next page

Table 3 *continued*

Country	National Policy and/or Action Plan	Year	Air Quality Targets	Sectoral Actions	AQ-Climate Integration	Haze/AP Episode Protocol
Viet Nam	• **Law on Environment Protection 2014** (revised 2020)	2020	✖	✔	✔	✖
	• QCVN 05:2023/BTNMT: National Technical Regulation on Air Quality (Revised $PM_{2.5}$ and PM_{10} Standards)	2023	✖	✖	✖	✖
	• Decision No. 19/2024/QD-Ttg: Roadmap for VES	2024	✖	✔	✖	✖
	• Circular No. 45/2024/TT-BTNMT: National Technical Regulation on Industrial Emissions	2024	✖	✔	✖	✖
	• Decision No. 876/QD-TTg: Approving the Action Program on Green Energy Transition in the Transport Sector	2022	✖	✔	✔	✖

ASEAN = Association of Southeast Asian Nations, AP = air pollution, AQ = air quality, DENR = Department of Environment and Natural Resources, Lao PDR = Lao People's Democratic Republic, PM = particulate matter, VES = vehicle emission standards.

Notes:

(i) A ✔ means the indicator is included in the policy or action plan, while a ✖ means it is not explicitly included.

(ii) Air Quality Targets—Pollution reduction goals with a target year (e.g. 41% $PM_{2.5}$ reduction by 2030).

(iii) Sectoral Actions—Defined actions across sectors (e.g., promote shift to nonmotorized transport).

(iv) Air Quality and Climate Change Integration—Integration of air quality and climate change strategies (e.g., decreased vehicle use = lower GHG emissions, better air quality).

(v) Haze/Air Pollution Episode Protocol—Structured response during episodes (e.g., Singapore's haze action plan).

(vi) Texts in bold refer to the primary air quality (Clean Air Acts) or environmental protection and management law.

Source: Clean Air Asia.

Subnational Implementation and Local Action

National authorities are encouraged to engage with their subnational counterparts more actively, recognizing their critical role in implementing localized solutions and sustaining long-term progress. While national legal frameworks provide the overarching structure for air quality management, subnational policies, plans, and strategies are essential for translating them into concrete, context-specific actions. These instruments allow provincial, city, and municipal governments to implement measures that respond to local pollution sources, geographic conditions, and population needs.

Seven ASEAN countries have adopted subnational policies or plans reflecting the distinct characteristics and priorities of their jurisdictions. No publicly accessible information was identified for Brunei Darussalam or Myanmar. In the case of Singapore, given its city-state structure, relevant instruments are considered at the national level. Most clean air action plans (CAAPs) or air quality management plans (AQMPs) include quantitative air pollutant reduction targets, sector-specific mitigation measures, and recognition of air quality and climate integration through a co-benefits approach. In the Philippines, Quezon City's "Executive Order No. 4, Series of 2023" established an Air Quality Management Technical Working Group composed of representatives from key city departments.[23] Part of the group's mandate is to develop, operationalize, and monitor the implementation of the Quezon City AQMP, finalized in the same year. In 2024, the city began piloting measures outlined in the plan. This example underscores how strong local leadership and interagency collaboration can drive meaningful progress on city-level air quality while complementing national-level goals.

Alignment with Climate Commitments

Lastly, ASEAN countries are encouraged to align their national and subnational air quality strategies with their nationally determined contributions (NDCs), to harness the co-benefits of climate and air pollution action. As presented in

[23] Quezon City Government, Philippines. 2023. Executive Order No. 4, Series of 2023: Creating the Quezon City Air Quality Management Technical Working Group. 13 February. https://quezoncity.gov.ph/public-notice/executive-order-no-4-series-of-2023-creating-thequezon-city-air-quality-management-technical-working-group.

Table 4, several countries have embedded air quality-related actions into their NDCs, such as promoting electric vehicles (EVs), shifting to renewable energy, and phasing down coal. These measures reduce GHG emissions while also improving ambient air quality. As countries update their NDCs, explicitly integrating air quality targets and indicators could strengthen alignment between climate and clean air objectives. Such integration supports more efficient policy implementation through cross-sector coordination and enhances overall environmental and public health outcomes. Leveraging synergies between air quality management and climate mitigation will be crucial to maximizing impact and mobilizing sustained investment for priority interventions across ASEAN.

Table 4: Subnational Air Quality Policies and Plans in ASEAN Countries

Country	Subnational Policy and/or Action Plans	Year	Air Quality Targets	Sectoral Actions	AQ-Climate Integration	Haze/AP Episode Protocol
Brunei Darussalam	• No recent major air quality policy identified	-	-	-	-	-
Cambodia	• Phnom Penh Sustainable City Plan 2018-2030	2019	✖	✔	✔	✖
Indonesia	• Jakarta Strategy for Air Pollution Control	2022	✔	✔	✔	✖
	• Bogor CAAP	2019	✖	✔	✔	✖
	• Tangerang CAAP (draft)	-	✔	✔	✔	✖
	• Jakarta's Governor Regulation 66/2020	2020	✖	✔	✖	✖
	• Jakarta's Clean Air Catalyst Initiative	2020	✖	✔	✔	✖
Lao PDR	• Green City Action Plan Vientiane Capital (2023-2030)	2024	✔	✔	✔	✖
Malaysia	• Kuala Lumpur Climate Action Plan, 2050	2021	✖	✔	✔	✖
Myanmar	• No recent major air quality policy identified.	-	-	-	-	-
Philippines	• Quezon City AQMP	2023	✔	✔	✔	✖
	• Iloilo City CAAP	2015	✖	✔	✔	✖
	• Parañaque City CAAP (2024–2026)	2024	✔	✔	✔	✖
	• Manila CAAP	2023	✔	✔	✔	✖
	• La Trinidad CAAP	2023	✔	✔	✔	✖
	• Marikina CAAP (2020–2030)	2019	✔	✔	✔	✖
	• Executive Order No. 4 Series of 2023: Creating the Quezon City Air Quality Management Technical Working Group	2023	✖	✖	✖	✖
Singapore	• No separate subnational plans, all air quality strategies are implemented nationally.	-	-	-	-	-
Thailand	• Chiang Mai Air Pollution Action Plan (2023–2030)	2023	✖	✔	✔	✖
	• Bangkok 16-Point Action Plan for $PM_{2.5}$ Management	2022	✖	✔	✖	✖
	• Bangkok Congestion Charge Plan (draft)	-	✖	✔	✔	✖
Viet Nam	• Ho Chi Minh City CAAP	2023	✔	✔	✔	✖
	• Ha Noi AQMP (2023–2030, Vision to 2035)	2024	✔	✔	✔	✖
	• Resolution No. 47/2024/NQ-HDND (Regulation on Low-Emission Zones in Ha Noi)	2024	✖	✔	✔	✖

ASEAN = Association of Southeast Asian Nations, AP = air pollution, AQ = air quality, AQMP = air quality management plan, CAAP = clean air action plan, PM = particulate matter, VES = vehicle emission standards.

Notes:

(i) The symbol ✔ means the indicator is included in the policy or action plan, while a ✖ means it is not explicitly included.

(ii) Air Quality Targets—Pollution reduction goals with a target year (e.g., 41% $PM_{2.5}$ reduction by 2030).

(iii) Sectoral Actions—Defined actions across sectors (e.g., promote shift to nonmotorized transport).

(iv) Air Quality and Climate Change Integration—Integration of air quality and climate change strategies (e.g., decrease vehicle use = lower greenhouse gas emissions, better air quality).

(v) Haze/Air Pollution Episode Protocol—Structured response during episodes (e.g., Singapore's haze action plan).

Source: Clean Air Asia.

Ambient Air Quality Standards in ASEAN

To better protect public health and strengthen air quality management, ASEAN countries are encouraged to adopt more stringent ambient air quality standards aligned with the latest WHO guidelines. Updating national standards reinforces the legal basis for pollution control and ensures that regulatory compliance leads to meaningful reductions in health risks.

Ambient air quality standards are established and monitored at national and/or city levels to protect populations from the adverse health effects of air pollution.[24] Governments typically draw from WHO guideline values while also considering national circumstances such as historical pollution levels, socioeconomic conditions, meteorology, topography, natural background concentrations, and exposed populations.[25] Publicly accessible information on ambient air quality standards is available for nine ASEAN countries, with Myanmar[26] currently in the process of establishing its national standards.[27]

Figure 17 shows that among the nine countries with a 24-hour ambient $PM_{2.5}$ standard, five are aligned with the less stringent WHO IT-3 (37.5 μg/m^3). Similarly, Figure 18 illustrates that five countries have annual $PM_{2.5}$ standards aligned with WHO IT-3 (15 μg/m^3). Malaysia and the Philippines have the most stringent short-term (24-hour) $PM_{2.5}$ standards at 35 μg/m^3, while Singapore has the most stringent long-term (annual) standard at 12 μg/m^3. However, these standards remain aligned with WHO IT-3 rather than the more protective WHO AQG or stricter interim targets. Other ASEAN countries have yet to update their national $PM_{2.5}$ standards in line with WHO AQG or to adopt more stringent interim targets. Given the well-established links between $PM_{2.5}$ exposure and adverse health outcomes, less stringent standards may create a false sense of safety when monitored concentrations are deemed compliant. Gradually strengthening ambient standards in line with updated WHO guidance would enhance public health protection and reinforce long-term policy ambition across the region.

Figure 17: National 24-Hour Ambient $PM_{2.5}$ Standards in ASEAN Countries Relative to WHO AQG

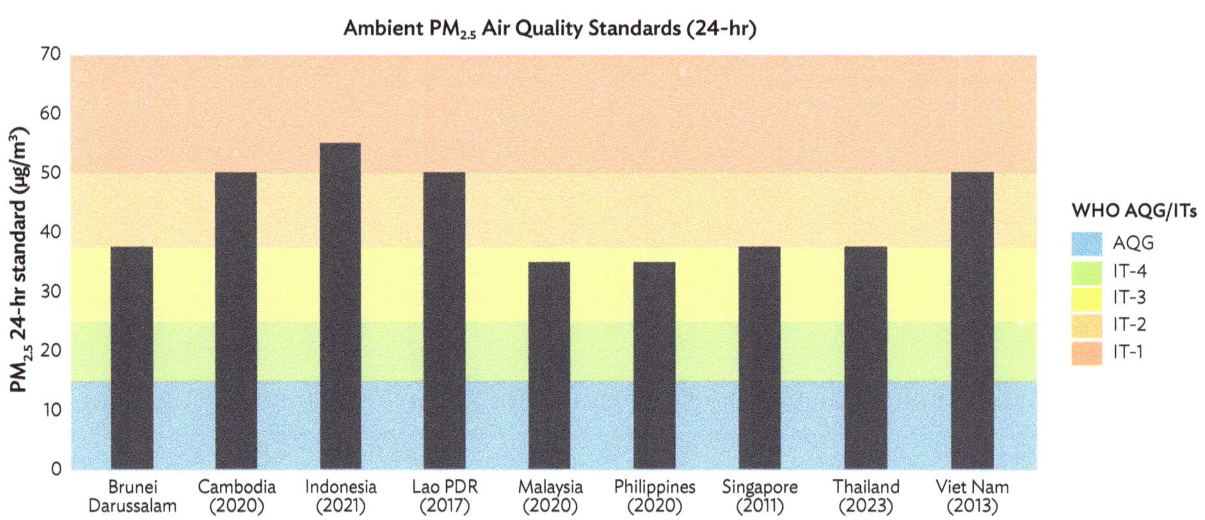

AQG = Air Quality Guidelines, ASEAN = Association of Southeast Asian Nations, IT = interim target, Lao PDR = Lao People's Democratic Republic, PM = particulate matter, WHO = World Health Organization.

Source: Clean Air Asia.

---

24 The terms "standards" and "guideline values" are commonly used interchangeably by most countries. This report adopted both terms to refer to the acceptable levels of air pollutants adopted by a national regulatory authority.

25 Clean Air Asia and United Nations Environment Programme (UNEP). 2019. Guidance Area 1: *Ambient Air Quality Standards and Monitoring*; Clean Air Asia. 2010. *Air Quality in Asia: Status and Trends 2010 Edition*.

26 UNEP. 2016. Air Quality Policies in Myanmar. UN Environment Programme.

27 Confirmation was obtained through email correspondence dated 31 July 2025 with a PhD student from Myanmar specializing in air quality research at the Asian Institute of Technology.

Figure 18: National Annual Ambient $PM_{2.5}$ Standards in ASEAN Countries Relative to WHO AQG

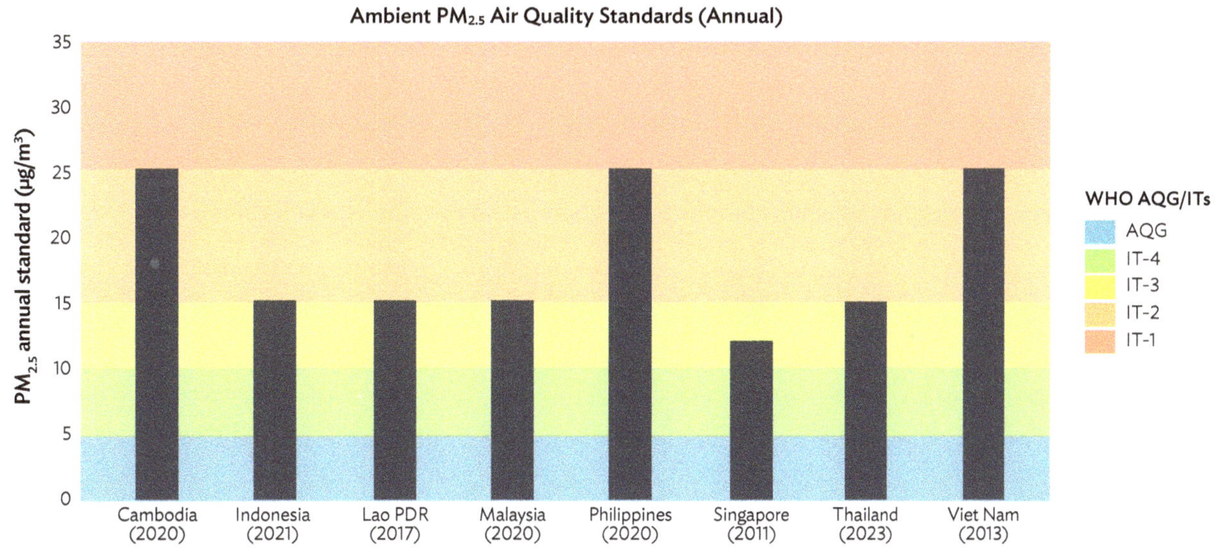

AQG = Air Quality Guidelines, ASEAN = Association of Southeast Asian Nations, IT = interim target, Lao PDR = Lao People's Democratic Republic, PM = particulate matter, WHO = World Health Organization.

Source: Clean Air Asia.

For PM_{10}, Figures 19 and 20 show that most 24-hour standards among ASEAN countries are aligned with WHO IT-1 (150 µg/m^3), while most annual standards are aligned with WHO IT-2 (50 µg/m^3). Singapore has adopted more stringent standards for both averaging periods, aligning with WHO IT-4 (50 µg/m^3 for 24-hour and 20 µg/m^3 for annual values). Indonesia aligns with IT-3 for its 24-hr standard, while Viet Nam and Malaysia align with IT-2. Overall, current PM_{10} standards in much of the region remain less protective than WHO AQG and more stringent interim targets. To complement efforts to progressively tighten $PM_{2.5}$ standards, ASEAN countries are encouraged to progressively adopt more protective PM_{10} standards.

Figure 19: National 24-Hour Ambient PM_{10} Standards in ASEAN Countries Relative to WHO AQG

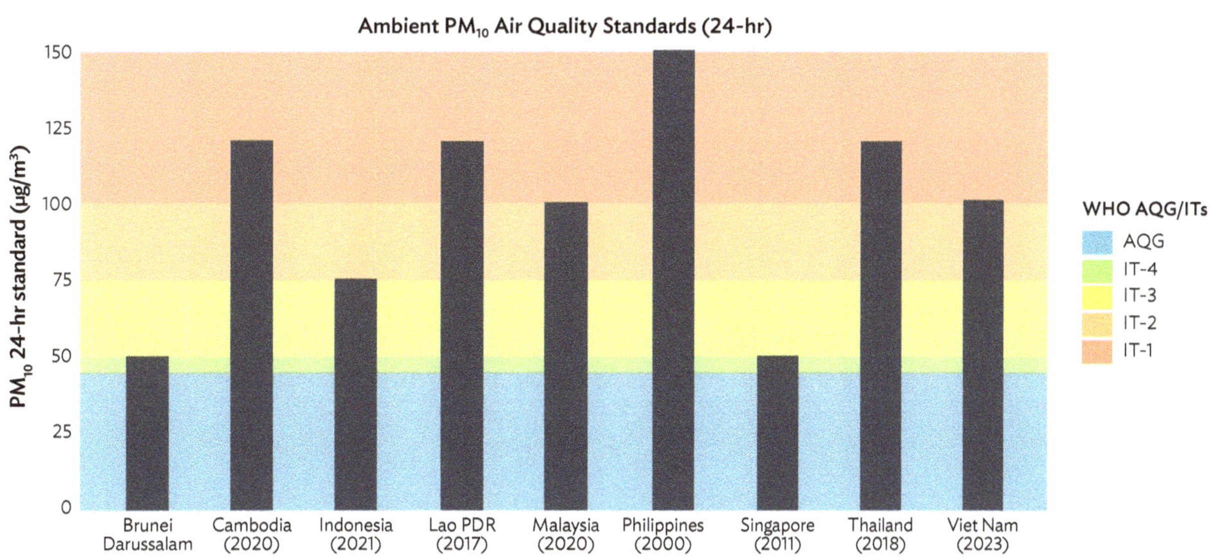

AQG = Air Quality Guidelines, ASEAN = Association of Southeast Asian Nations, IT = interim target, Lao PDR = Lao People's Democratic Republic, PM = particulate matter, WHO = World Health Organization.

Source: Clean Air Asia.

Figure 20: National Annual Ambient PM_{10} Standards in ASEAN Countries Relative to WHO AQG

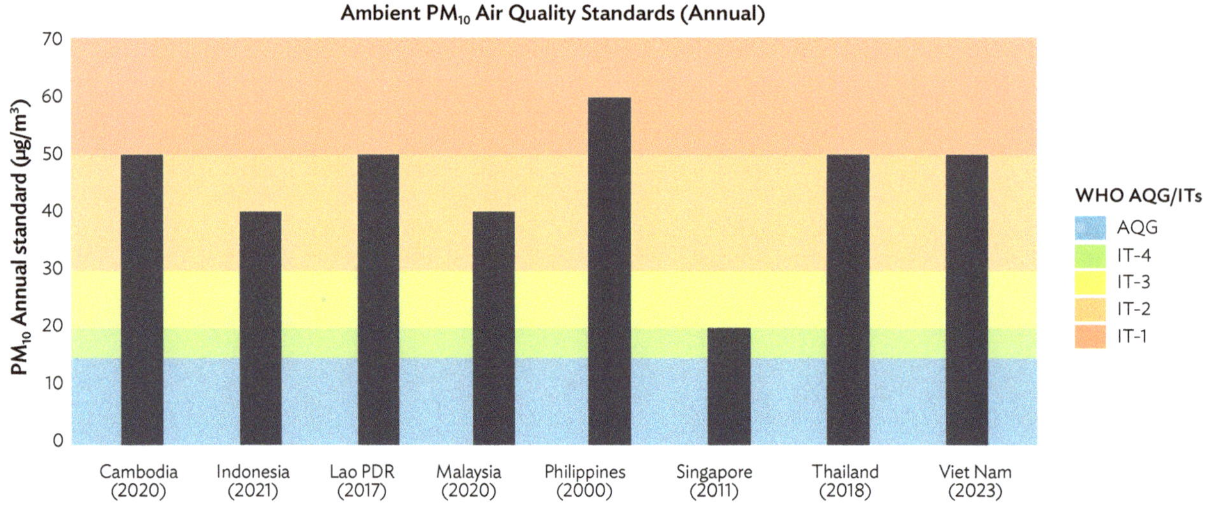

AQG = Air Quality Guidelines, ASEAN = Association of Southeast Asian Nations, IT = interim target, Lao PDR = Lao People's Democratic Republic, PM = particulate matter, WHO = World Health Organization.

Source: Clean Air Asia.

For nitrogen dioxide (NO_2), Figures 21 and 22 show that only a limited number of ASEAN countries have established national standards. Six countries have 24-hr standards and six have annual standards, although these do not necessarily overlap. For both averaging periods, existing standards correspond to WHO IT-1 (24-hour: 120 µg/m³; annual: 40 µg/m³), the least stringent level. Indonesia has the most stringent 24-hour standard at 65 µg/m³, while the Lao PDR has the most stringent annual standard at 37.6 µg/m³. However, both remain within WHO IT-1 range. In several cases, national standards exceed the IT-1 threshold for one or both averaging periods. The relatively limited coverage and modest ambition of NO_2 standards across the region suggest scope for further revision. As NO_2 serves as a primary indicator of combustion-related emissions, particularly from transport and industry, more protective national limits would enhance regulatory credibility and improve control of urban emission sources.

Figure 21: National 24-Hour Ambient NO_2 Standards in ASEAN Countries Relative to WHO AQG

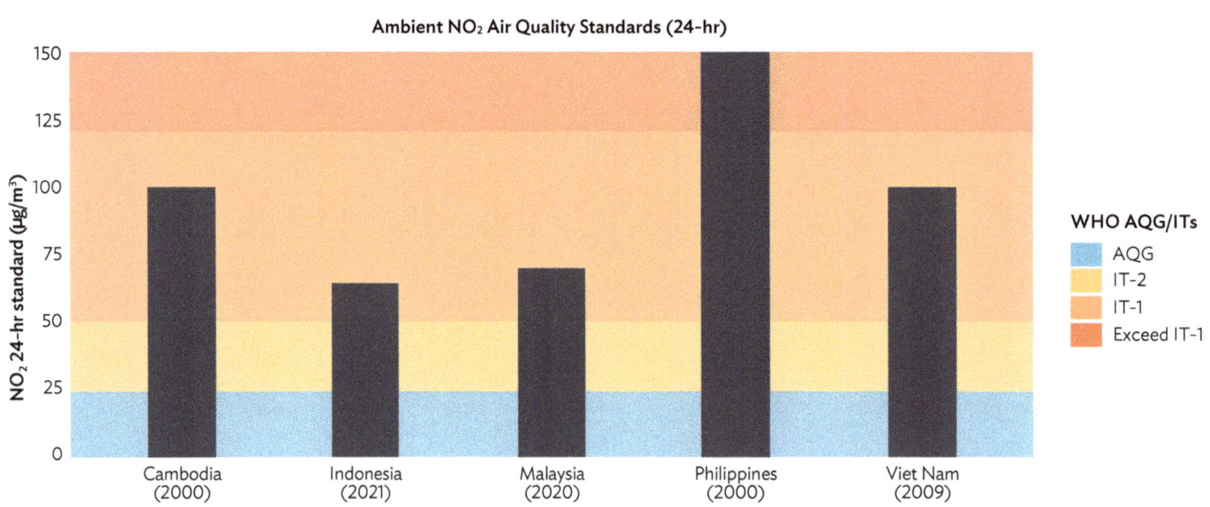

AQG = Air Quality Guidelines, ASEAN = Association of Southeast Asian Nations, IT = interim target, NO_2 = nitrogen dioxide, WHO = World Health Organization.

Source: Clean Air Asia.

Figure 22: National Annual Ambient NO_2 Standards in ASEAN Countries Relative to WHO AQG

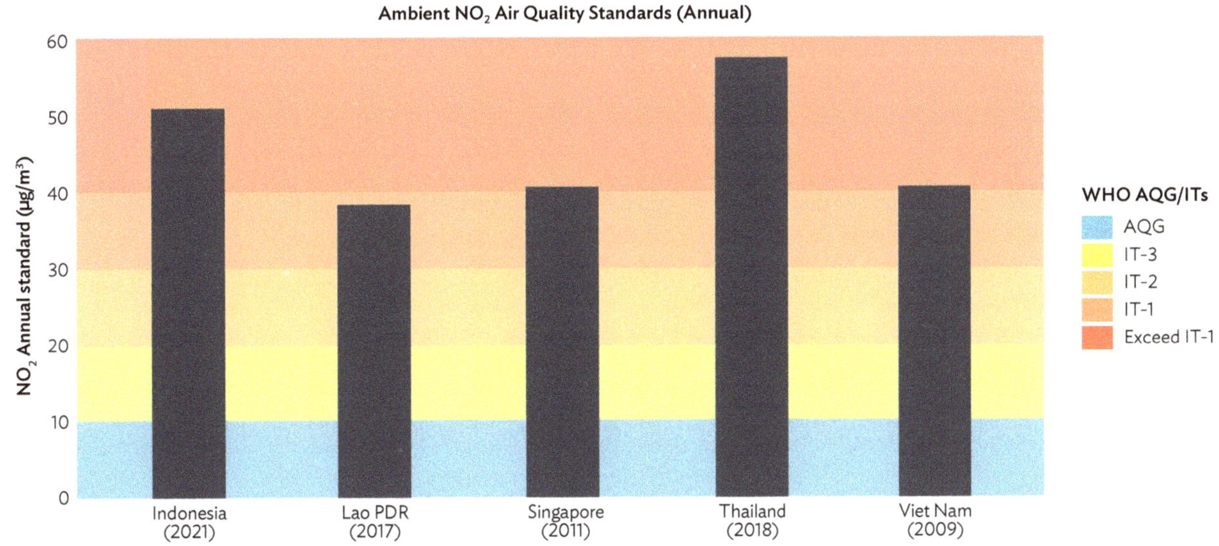

AQG = Air Quality Guidelines, ASEAN = Association of Southeast Asian Nations, IT = interim target, Lao PDR = Lao People's Democratic Republic, NO_2 = nitrogen dioxide, WHO = World Health Organization.

Source: Clean Air Asia.

The full set of national ambient air quality standards across pollutants is presented in Table 5.

Table 5: National Ambient Air Quality Standards in ASEAN Countries

Country	$PM_{2.5}$		PM_{10}		TSP		SO_2			NO_2			O_3		CO	
	24-Hour	Annual	24-Hour	Annual	24-Hour	Annual	1-Hour	24-Hour	Annual	1-Hour	24-Hour	Annual	1-Hour	8-Hour	1-Hour	8-Hour
Brunei Darussalam	37.5	-	50	-	-	-	-	50	-	100	-	-	-	100	30	10
Cambodia	50	25	120	50	330	100	500	300	100	300	100	-	200	-	40	20
Indonesia	55	15	75	40	230	-	150	75	45	200	65	50	150	100	10	4
Lao PDR	50	15	120	50	330	50	340.6	131	-	206.8	-	37.6	200	140	34.5	10.35
Malaysia	35	15	100	40	-	-	250	80	-	280	70	-	180	100	30	10
Myanmar	-	-	-	-	-	-	-	-	-	-	-	-	-	-	-	-
Philippines	35	25	150	60	230	90	-	180	80	-	150	-	140	60	35	10
Singapore	37.5	12	50	20	-	-	-	50	-	200	-	40	-	100	30	10
Thailand	37.5	15	120	50	330	100	780	300	100	320	-	57	200	140	34.2	10.26
Viet Nam	50 (45)[a]	25	100	50	200	100	350	125	50	200	100	40	200	120	30	10

ASEAN = Association of Southeast Asian Nations, CO = carbon monoxide, Lao PDR = Lao People's Democratic Republic, NO_2 = nitrogen dioxide, O_3 = ground-level ozone, PM = particulate matter, SO_2 = sulfur dioxide, TSP = total suspended particulates.

Note: In µg/m^3, unless otherwise indicated. CO is in mg/m^3.

[a] Viet Nam's updated 24-hour $PM_{2.5}$ standard will be implemented in 2026.

Source: Clean Air Asia.

Air Quality in Public Health Management

Air pollution is a significant risk factor for disease and premature mortality across ASEAN, underscoring the importance of embedding air quality considerations within public health planning. Integrating air quality management into national health strategies could take several forms, including institutionalized exposure risk assessments, mapping of health and economic impacts, and the incorporation of air quality indicators into surveillance systems and early warning mechanisms. Seven countries have adopted national health action plans or policies that explicitly recognize the health impacts of poor air quality, as listed below.

1. Cambodia: National Climate Change Action Plan for Public Health, 2019–2023
2. Lao People's Democratic Republic: Health National Adaptation Plan
3. Malaysia: National Environmental Health Action Plan
4. Philippines: National Environmental Health Action Plan, 2030
5. Singapore: Healthy Living Master Plan
6. Thailand: National Environmental Health Action Plan
7. Viet Nam: National Action Plan for the Health Sector Response to Climate Change (2019–2030, with a vision to 2050)

These plans create an opportunity to address the adverse health effects of air pollution through closer collaboration between environmental and health authorities, enabling more coordinated resource allocation and integrated policy design. Where similar national-level policies are not yet in place, subnational initiatives—such as the Central Kalimantan Health and Pollution Action Plan in Indonesia—can provide a foundation for scaling up action nationwide.

According to the *State of Global Air 2025*, the Lao PDR and Myanmar recorded the highest mortality rates (per 100,000 population) related to air pollution in the region, as shown in Figure 23. This highlights the importance of formalizing institutional linkages between air quality management and public health systems. Leveraging existing health strategies to address air pollution can improve cross-sectoral alignment, support more targeted interventions, and contribute to reducing pollution-related disease burdens across the ASEAN region.

Figure 23: Air Pollution Mortality Rate in ASEAN Countries and the Presence of Air Pollution-Related Policies and Plans

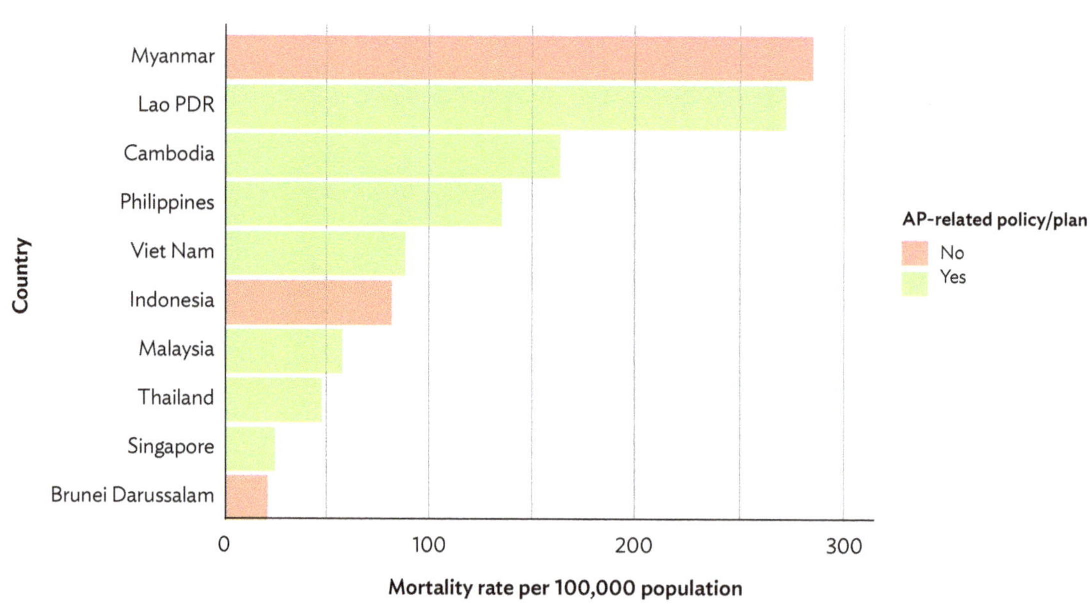

AP = air pollution, ASEAN = Association of Southeast Asian Nations, Lao PDR = Lao People's Democratic Republic.
Note: Mortality data refer to air pollution–attributable mortality rates per 100,000 population, based on estimates from the Global Burden of Disease Study 2023, as reported in the State of Global Air 2025. Data reflect 2023, the most recent year available.
Source: Health Effects Institute. 2025. State of Global Air 2025.

Sectoral Interventions

In response to identified emission drivers, several ASEAN countries have introduced sector-specific measures, particularly in transport and energy, which are major sources of urban air pollution. As summarized in Tables 6 to 11, Indonesia, Malaysia, the Philippines, Singapore, Thailand, and Viet Nam have established targets to accelerate electric vehicle (EV) adoption. These initiatives range from electrification of public transport fleets to broader nationwide EV deployment strategies, often supported by charging infrastructure expansion and government fleet procurement programs.

Table 6: E-Mobility Targets in Indonesia

Year	Production/Construction Target
2025	• EV cars: 2,200 • EV motorcycles: 2,130,000 • Charging stations: 1,000
2050	• Cars: 4,200,000 • Motorcycles: 13,300,000 • Charging stations: 10,000

EV = electric vehicle.
Source: Government of Indonesia, Ministry of Energy and Mineral Resources. 2020. Govt to Speed Up EV Infrastructure Construction. News release. 4 October.

Table 7: E-Mobility Targets in Malaysia

Year	Action Plans
2021–2030	• Implement EV procurement for government fleet. • Provide incentives for EV purchase of taxis and income tax exemptions to offset high capital cost of EVs.
2021–2025	• Create additional scoring for local content.
2022	• 10% target of new government fleet to be BEVs.
2023–2025	• 50% target of new government fleet to be BEVs. • 20% target of new government-linked companies' fleet to be BEVs.
2025–2030	• Establish local product qualification for tender participation. • Establish 50% target of new government fleet to be locally manufactured BEVs.

BEV = battery electric vehicle, EV = electric vehicle.
Source: Government of Malaysia, Ministry of Environment and Water. 2021. *Low Carbon Mobility Blueprint, 2021–2030.*

Table 8: E-Mobility Targets in the Philippines

Sales Targets	Business-as-Usual Scenario			Clean Energy Scenario		
	Short Term (2023–2028)	Medium Term (2029–2034)	Long Term (2035–2040)	Short Term (2023–2028)	Medium Term (2029–2034)	Long Term (2035–2040)
Cars						
Hybrid (HEV)	81,500	49,000	36,600	415,000	234,000	107,000
Plug-In Hybrid (PHEV)	13,600	24,600	36,600	69,000	80,000	107,000
BEV	13,600	123,000	219,400	69,000	327,000	641,000
Tricycles (BEV)	37,500	71,000	103,400	419,000	262,000	223,000
Motorcycles (BEV)	164,900	311,800	454,400	1,800,000	947,000	922,000
Buses (BEV)	600	1,200	1,800	2,200	1,500	1,600
EV Charging Stations	7,300	14,000	20,400	66,500	41,800	39,800

BEV = battery electric vehicle, HEV = hybrid electric vehicle, PHEV = plug-in hybrid electric vehicle.
Source: Government of the Philippines, Department of Energy. 2023. *Comprehensive Roadmap for Electric Vehicle Industry.*

Table 9: E-Mobility Targets in Singapore

Year	Target
2025	• 400 diesel buses to be replaced with e-buses (60 buses already deployed as of end 2021).
2030	• 60,000 electric vehicle charging points (40,000 in public carparks and 20,000 in private premises). • Electrification of half the public bus and taxi fleet.
2040	• 100% of vehicles to run on cleaner energy.

Source: Government of Singapore. 2021. Singapore Green Plan 2030.

Table 10: E-Mobility Targets in Thailand

Phase	Manufacturing Target
1 (2021–2022)	• Promote e-motorcycle and support infrastructure.
2 (2023–2025)	• Achieve economies of scale through the production of » 225,000 cars and pickup trucks, » 360,000 motorcycles, and » 18,000 buses or trucks.
3 (2026–2030)	• Produce: » 725,000 EV cars and pickup trucks, and » 675,000 EV motorcycles including batteries.

EV = electric vehicle.
Source: National Electric Vehicle Policy Committee. 2021. *30@30 Policy.*

Table 11: E-Mobility Targets in Viet Nam

Year	Target
2022–2030	• Encourage the production, importation, assembly, and utilization of EVs. • Develop EV charging infrastructure.
2025	• 100% of buses to use electricity and green energy.
2030	• 100% of taxis to use electricity and green energy. • 50% of vehicles to use electricity and green energy.
2040	• Discontinue production, assembly, and import of fossil fuel vehicles, motorcycles, and mopeds for domestic use.
2050	• Achieve net-zero greenhouse gas emissions from the transport sector. • Complete green transport network (100% road vehicles, heavy construction vehicles and/or equipment, buses, and taxis to use electricity and green energy; bus stations and rest stops to meet green criteria).

EV = electric vehicle.
Source: Government of Viet Nam. 2022. Decision No. 876/QD-TTg–Approving the Action Program for Transition to Green Energy and Mitigation of Carbon Dioxide and Methane Emissions from Transportation.

The region has also made notable advances in vehicle emission and fuel quality standards. As illustrated in Figure 24, most ASEAN members are implementing Euro IV emission standards and are expected to maintain these requirements for applicable vehicle categories through 2030. While this marks improvement compared to earlier regulatory levels, further reductions in transport-related emissions could be achieved through the adoption of more advanced standards. Singapore leads in this area, having fully adopted the Euro VI standards for both petrol and diesel vehicles. Regarding fuel quality, as of 2025, half of the ASEAN members have adopted a maximum sulfur content of 10 parts per million for diesel and gasoline fuels. In contrast, Figure 25 shows that sulfur limits remain significantly higher in some jurisdictions. Indonesia, for example, permits up to 500 parts per million for diesel and 350 parts per million for petroleum. Differences in fuel sulfur levels influence the effectiveness of vehicle emission control technologies and therefore affect overall air quality outcomes.

Figure 24: Adoption of Euro-Equivalent Vehicle Emission Standards in ASEAN Countries

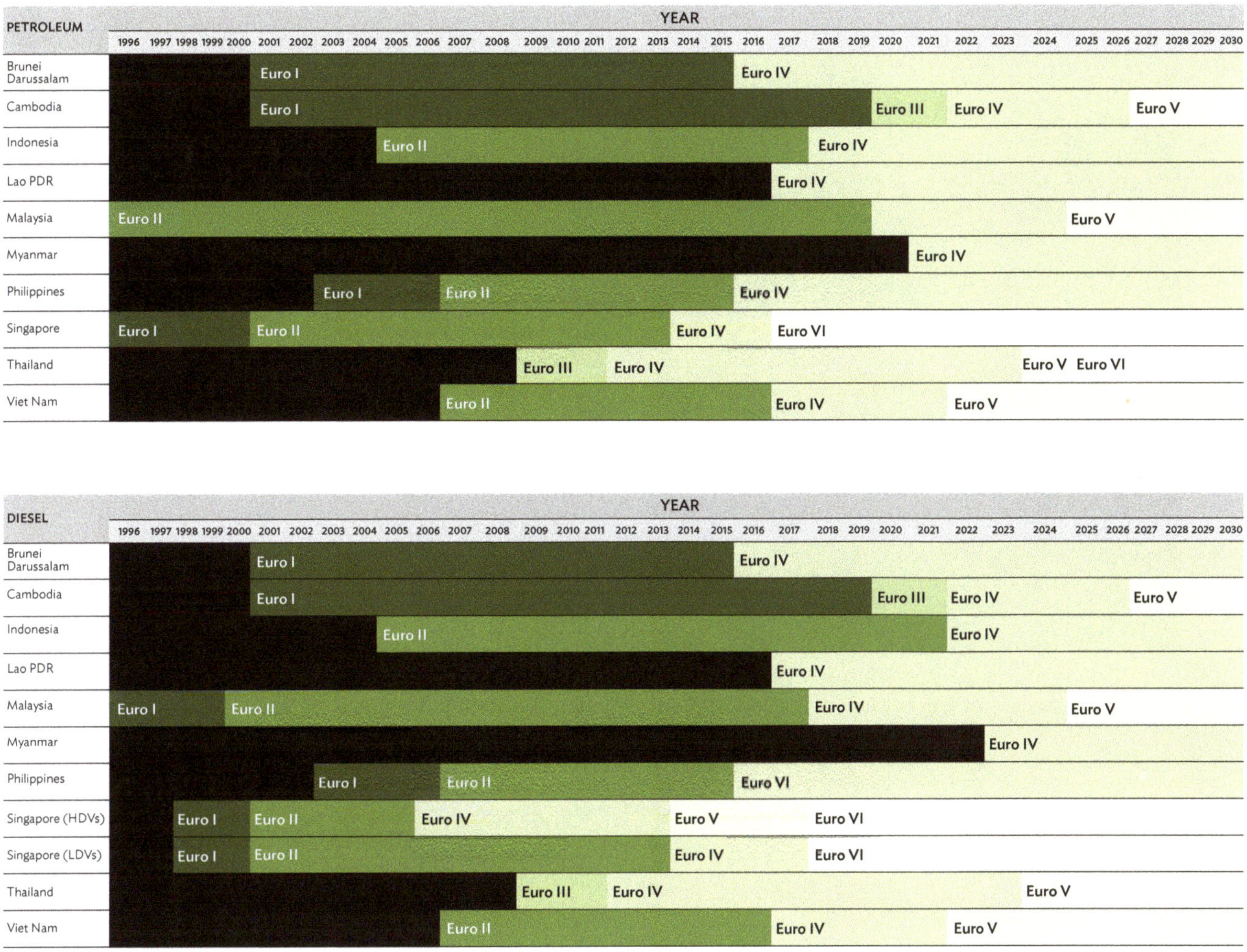

ASEAN = Association of Southeast Asian Nations, HDV = heavy-duty vehicles, Lao PDR = Lao People's Democratic Republic, LDV = light-duty vehicles.

Note: Lighter shades of green indicate higher Euro standards and cleaner fuel quality, while darker shades represent earlier or less stringent standards.

Source: Clean Air Asia. 2026. Current and Proposed Diesel Vehicle Emission Standards in Asia.

Figure 25: Sulfur Content Standards for Diesel and Petroleum in ASEAN Countries

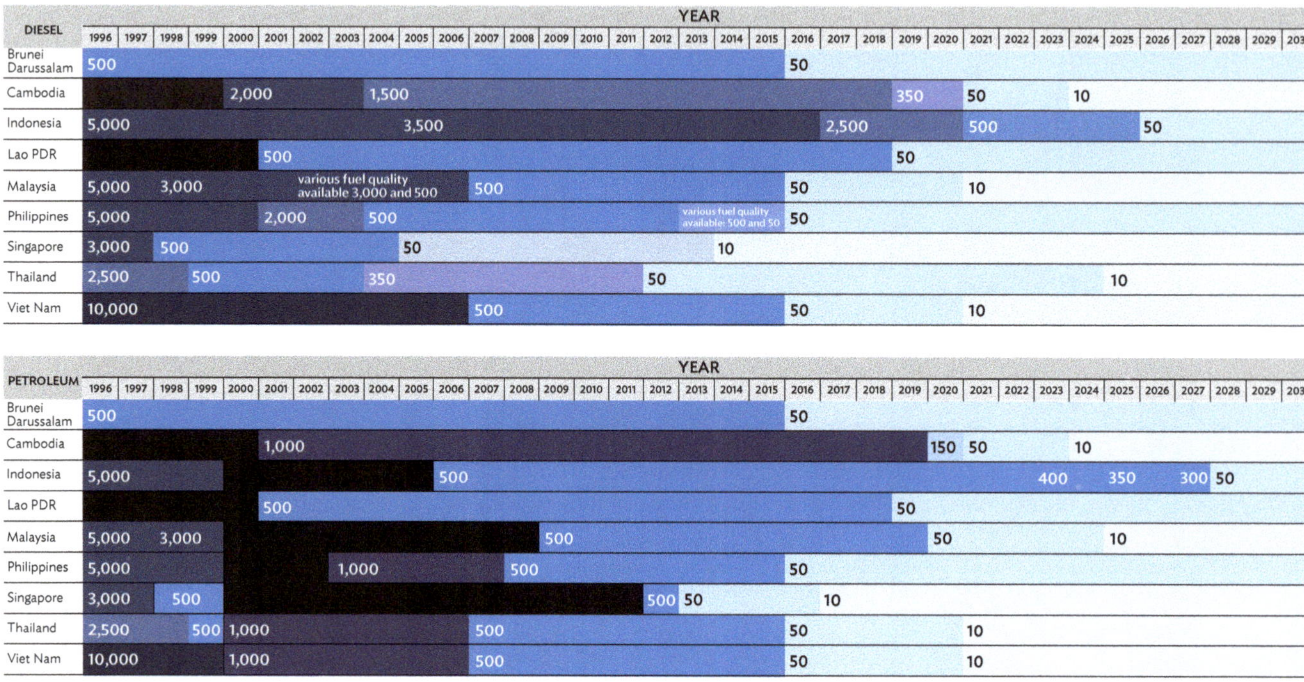

ASEAN = Association of Southeast Asian Nations, Lao PDR = Lao People's Democratic Republic.
Note: Values are expressed in parts per million and represent the maximum sulfur content allowed in fuels where lower values (shown in lighter shades of blue) indicate higher fuel quality and alignment with stricter Euro emission standards; black indicates no regulation. For the Lao PDR (diesel and petroleum), the implementation date of the latest declared standard is uncertain. For Malaysia, the Philippines, and Viet Nam (diesel), the latest standards are mandatory under national policy.
Source: Clean Air Asia. January 2026. Current and Proposed Petrol Sulfur Levels in Asia.

Coal continues to play a significant role in electricity generation in several ASEAN countries and remains an important source of particulate matter (PM) and gaseous pollutants. Emission limits for PM, nitrogen oxides (NO_x), and sulfur oxides (SO_x) from coal-fired power plants (CFPs) in parts of the region, as presented in Figures 26 to 28, remain comparatively lenient[28] compared to those in jurisdictions such as the People's Republic of China and the European Union. In some cases, allowable emissions thresholds exceed international best practice levels by a considerable margin. Revisiting these limits offers potential to reduce power sector contributions to ambient air pollution concentrations. At the same time, several countries are beginning to moderate coal dependence through broader energy transition strategies as shown in Table 12. Indonesia and Viet Nam have introduced policies to phase down or restrict new coal developments and expand renewable energy deployment under their nationally determined contributions (NDCs). Other countries, including the Philippines, have not formalized coal phaseout timelines but are gradually diversifying their energy mix toward lower-emissions sources.

[28] Clean Air Asia. 2020. *Coal-Fired Power Plant Emission Standards in South and Southeast Asian Countries: Policy Analysis.*

Figure 26: PM Standards for Coal-Fired Power Plants in ASEAN Countries

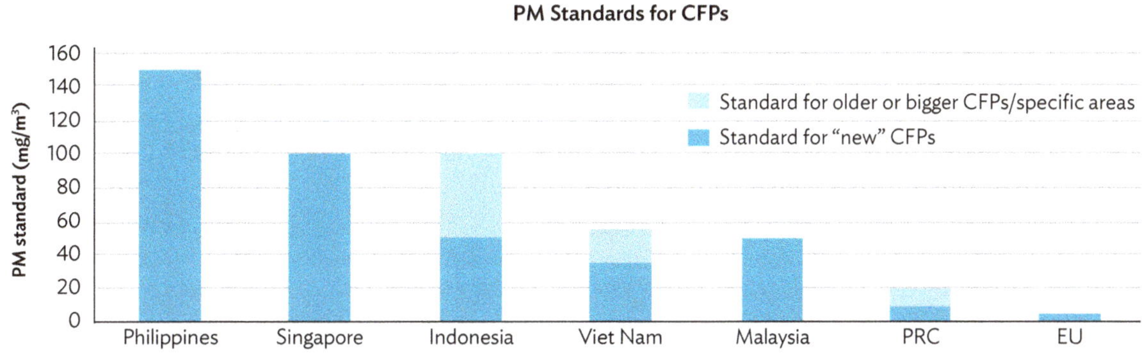

ASEAN = Association of Southeast Asian Nations, CFP = coal-fired power plants, EU = European Union, PM = particulate matter, PRC = People's Republic of China.

Source: Clean Air Asia.

Figure 27: NO_X Standards for Coal-Fired Power Plants in ASEAN Countries

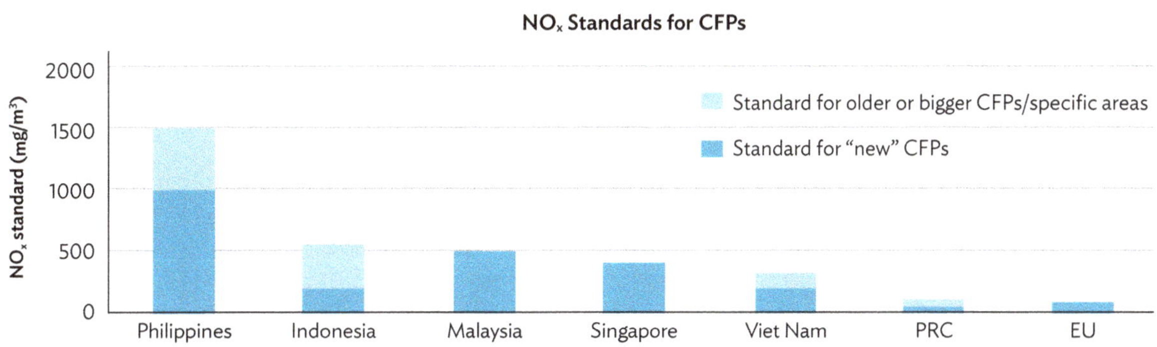

ASEAN = Association of Southeast Asian Nations, CFP = coal-fired power plants, EU = European Union, NO_X = nitrogen oxides, PRC = People's Republic of China.

Source: Clean Air Asia.

Figure 28: SO_x Standards for Coal-Fired Power Plants in ASEAN Countries

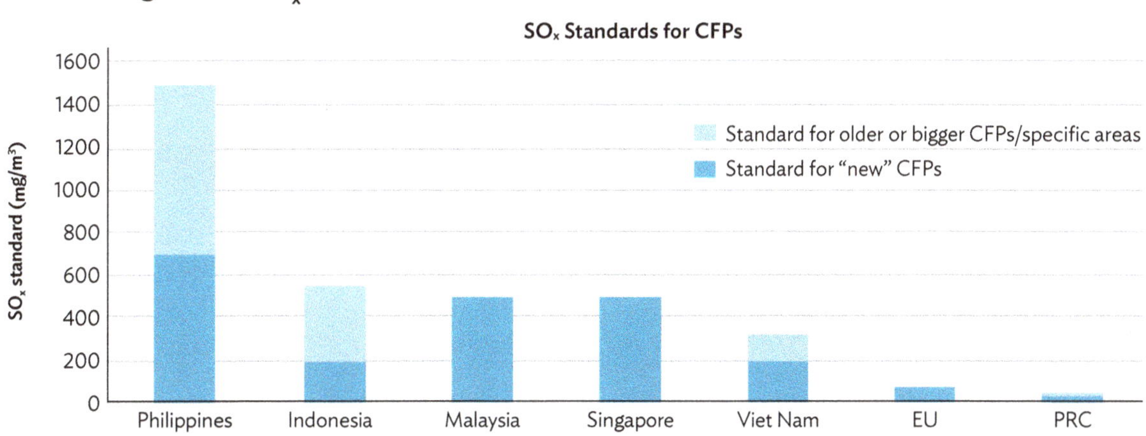

ASEAN = Association of Southeast Asian Nations, CFP = coal-fired power plants, EU = European Union, PRC = People's Republic of China, SO_x = sulfur oxides.

Source: Clean Air Asia.

Table 12: Nationally Determined Contributions and National Policies and Strategies Related to Coal Use in ASEAN Countries

Country	NDC Targets	Policies to Meet NDC Targets Related to Coal Use
Cambodia	Reduction in emissions by 41.7% by 2030 compared to BAU[a]	The Long-Term Strategy for Carbon Neutrality sets out the following mitigation actions: • No new coal generation capacity beyond already committed projects • Substitution of coal in the industrial and power sector[b]
Indonesia	GHG emission reduction by 2030 compared to BAU: By 31.89% (unconditional) By 43.2% (conditional)[c]	Presidential Regulation No. 112 of 2022 bans new CFPs, with exceptions for those in the existing power development plan and mandates a roadmap for early plant retirement.[d] Perusahaan Listrik Negara (PLN) has outlined plans to retire all CFPs by 2056 and halt the approval of new ones.[e]
Lao People's Democratic Republic	Reduce GHG emission by 60% by 2030 (unconditional) compared to the baseline scenario[f]	The Draft National Power Development Plan (NPDP) 2020–2030 indicates increasing coal utilization for power generation.
Malaysia	Reduction in carbon intensity against gross domestic product (GDP) by 45% by 2030 relative to 2005 levels (unconditional)[g]	By 2050, the government targets to phase out coal as an energy source and 70% renewable energy installed capacity is achieved. Natural gas is expected to become a transition fuel from baseload coal.[h]
Philippines	Reduce GHG emissions by 75% from the BAU scenario by 2030 for the sectors of agriculture, wastes, industry, transport, and energy, of which 2.71% is unconditional and 72.29% is conditional[i]	There are no phaseout strategies or policies at the national level, and the Philippine Energy Plan (PEP) does not include a road map for coal phase-out. However, the Philippine government is encouraging a voluntary early and orderly decommissioning or repurposing of existing CFPs (phase down), as demonstrated in the PEP, which shows a decrease in reliance on coal for power generation. Efforts are being made to secure a stable supply and address the climate emergency by ramping up the country's renewable energy target of more than 50% share by 2050.[j]
Thailand	Reduction in GHG emissions by 2030 compared to BAU: By 30% (unconditional) By 40% (conditional)[k]	The updated Power Development Plan indicates that the percentage of power generating capacity from coal will decrease from 12.8% in 2018 to 6.3% in 2037, while the percentage from domestic renewable energy sources will increase from 13.5% in 2018 to 32.5% in 2037. Natural gas will account for 42.5% of the power generating capacity in 2037.[l]
Viet Nam	GHG emission reduction by 2030 compared to BAU: By 15.8% (unconditional) By 43.5% (conditional)[m]	Eighth National Power Development Plan (PDP8) (2021–2030) • Viet Nam will stop building new coal power plants after 2030 After 20 years of operation, CFPs will burn biomass fuel, starting at 20% and gradually increasing to 100%. CFPs with operational lifespans over 40 years will be shut down if fuel conversion is not possible.[n]

ASEAN = Association of Southeast Asian Nations, BAU = business-as-usual, CFP = coal-fired power plant, GHG = greenhouse gas, NDCs = nationally determined contributions.

[a] Government of Cambodia, National Council for Sustainable Development. 2020. *Cambodia's Updated Nationally Determined Contribution.*
[b] Government of Cambodia, National Council for Sustainable Development. 2021. *Long-Term Strategy for Carbon Neutrality.*
[c] Republic of Indonesia. 2022. *Enhanced Nationally Determined Contribution.*
[d] Ministry of Finance, Republic of Indonesia. 2022. Presidential Regulation of the Republic of Indonesia, Number 112 of 2022.
[e] ASEAN Centre for Energy. 2024. *Assessment of the Role of Coal in the ASEAN Energy Transition and Coal Phase-Out.*
[f] Government of the Lao People's Democratic Republic. 2021. *Nationally Determined Contribution.*
[g] Government of Malaysia. 2021. Malaysia's Update of its *First Nationally Determined Contribution.*
[h] Ministry of Economy, Government of Malaysia. n. d. *National Energy Transition Roadmap.*
[i] Government of the Philippines. 2021. *Nationally Determined Contribution.*
[j] Department of Energy, Government of the Philippines. 2024. *Philippine Energy Plan, 2023–2050.*
[k] Government of Thailand. 2022. *Thailand's Second Updated Nationally Determined Contribution.*
[l] Ministry of Energy, Government of Thailand. 2018. *Power Development Plan, 2018–2037.*
[m] Government of Viet Nam. 2022. *Nationally Determined Contribution.*
[n] Government of Viet Nam. 2025. *Adjustment of the National Power Development Plan for the 2021–2030 Period, with a Vision to 2050 (PDP8).*

Source: This table was first published by Clean Air Asia. 2020. *Coal-Fired Power Plant Emission Standards in South and Southeast Asian Countries: Policy Analysis.*

Public Reporting

Clear and consistent communication of air quality information is essential for translating technical data into practical guidance for the public. ASEAN countries can enhance the effectiveness of air quality reporting by periodically reviewing existing AQI systems and considering approaches that more directly reflect health risk. Improving how air quality is conveyed helps empower people to take informed action and reinforces the link between environmental monitoring and health awareness.

The AQI is widely used to communicate air pollution levels in formats accessible to the general public. It relies on a unitless numerical scale, where higher index values indicate poorer air quality. Index values are derived from pollutant concentration ranges, known as "breakpoints," which vary by country and pollutant. Since there is no single international methodology for calculating AQIs, national systems vary. AQIs typically combine numerical values with descriptive labels such as "good," "fair," "unhealthy," and "hazardous," and corresponding color codes, such as green or blue for good air quality and red or darker shades for more polluted conditions. Each category is accompanied by health advisories intended to guide behavior during pollution episodes.

To assess regional differences, Figure 29 compiles national $PM_{2.5}$ breakpoints, given the strong association between $PM_{2.5}$ exposure and adverse health outcomes. The comparison revealed considerable variation across ASEAN. As a result, identical $PM_{2.5}$ concentrations may be classified under different AQI categories depending on the country. In the Philippines, $PM_{2.5}$ levels between 0.0 to 25.0 $\mu g/m^3$ are still considered "good," while in Singapore, a value of 13 $\mu g/m^3$ would already be categorized as "moderate." These differences reflect national policy choices, local environmental conditions, and health-based regulatory thresholds. Periodic review of AQI breakpoints can help ensure that communication tools remain aligned with evolving scientific evidence and public health objectives.

To better integrate air quality and health management, some countries have adopted or are evaluating the use of an Air Quality and Health Index (AQHI). Unlike traditional AQIs, which are based on concentration breakpoints, the AQHI incorporates epidemiological data to estimate short-term health risk from exposure to multiple pollutants. This risk-based approach provides more nuanced guidance, especially for vulnerable populations, and can enhance decision-making during high pollution events.

Dalat city downtown aerial view in Viet Nam (photo by ADB).

Figure 29: $PM_{2.5}$ Breakpoints by Air Quality Index in ASEAN Countries

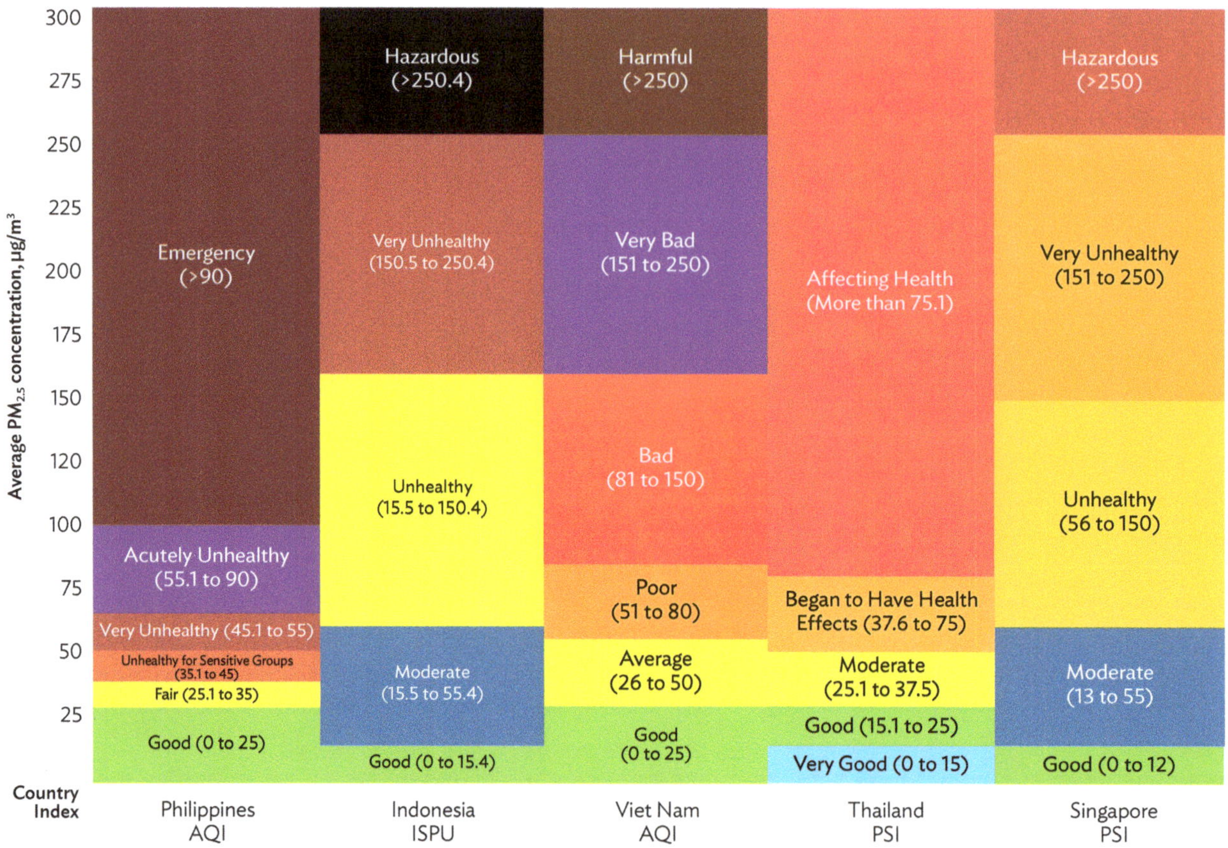

Note: The numbers in parenthesis correspond to the breakpoint values.

AQI = Air Quality Index, ASEAN = Association of Southeast Asian Nations, ISPU = Indonesia's Air Pollution Standard Index, PM = particulate matter, PSI = Pollutant Standards Index.

Source: Clean Air Asia.

5 Financing Clean Air Action

Effective governance frameworks, standards, sectoral strategies, and public reporting systems establish the foundation for improving air quality. However, translating policy ambition into measurable emission reductions ultimately depends on the availability of sustained and well-structured financing. Scaling clean air interventions across the ASEAN region will require mobilizing resources commensurate with the scale of the challenge.

Air quality solutions remain significantly underfunded, receiving only 1% of international development finance.[29] Domestic allocations for clean air are also limited. Across all eight cities examined in this report, financing gaps persist in sectoral investments that could deliver substantial reductions in emissions. In addition, many cities face constraints in developing bankable project pipelines, preparing investment-ready proposals, and accessing appropriate financing instruments.

Clean air interventions vary widely in scope and scale, ranging from policy reforms and monitoring upgrades to capital-intensive infrastructure investments in transport, energy, waste management, and industry. As such, a diversified mix of financing modalities is required. No single instrument can address the full spectrum of investment needs. Financing mechanisms can broadly be grouped into public finance, private finance, and catalytic instruments designed to mobilize additional capital. Potential financial mechanisms include the following:[30]

- **Grants (public or philanthropic).** Nonrepayable financing typically used to support policy reform, technical assistance, capacity building, emissions inventories, monitoring system upgrades, feasibility studies, and early-stage preparation of clear air investment projects.
- **Debt finance (loans and bonds).** Repayable financing that can either support policy or institutional reforms (through policy-based lending) or capital-intensive infrastructure investments such as public transport electrification, industrial emission control retrofits, waste treatment facilities, and clean energy systems. In some cases, loans may incorporate concessional terms, such as reduced interest rates or longer repayment periods, or be delivered through financial intermediaries to incentivize small and medium-sized enterprises and technology providers to adopt cleaner production systems.
- **Guarantees and risk-mitigation instruments.** Mechanisms in which a third party shares or absorbs part of the financial risk to unlock private capital, improve creditworthiness, and enable investment in areas such as pollution control technologies, clean fuel switching, large-scale urban transport, or waste infrastructure projects.
- **Results-based finance.** Financing in which disbursements are linked to verified performance outcomes, such as achieved emissions reductions,[31] compliance improvements in industrial facilities, or implementation milestones.
- **Blended finance.** The strategic use of concessional capital to de-risk projects, enhance bankability, and crowd in additional public and private investment in clean air interventions, including electric mobility systems, industrial retrofits, and clean energy deployment.
- **Equity finance.** Direct investment in companies or special purpose vehicles developing or deploying clean technologies, including electric mobility, clean fuels, pollution monitoring systems, and industrial innovation. Equity instruments can help scale emerging solutions where traditional debt financing may be constrained.
- **Market-based and fiscal instruments.** Policy tools such as pollution charges, emissions trading systems, fuel taxes, and congestion pricing mechanisms that both incentivize emission reductions and generate revenue that can be reinvested in public transport upgrades, monitoring systems, and other clean air programs.

---

29 Clean Air Fund. 2025. *The State of Global Air Quality Funding 2025*.

30 Based on the *Climate and Clean Air Coalition's Clean Air Flagship's Financing for Clean Air: Landscape Review.*

31 Asian Development Bank. 2021. *What is Results-Based Lending?*

A notable example of deploying multiple financial instruments at scale is the Air Quality Improvement Program in the greater Beijing–Tianjin–Hebei region in the People's Republic of China. Implementing the region's clean air action plan required investment far beyond available public resources,[32] necessitating a coordinated, multi-instrument financing approach.

The government, in partnership with multilateral development institutions and other financiers, including the Asian Development Bank (ADB), adopted a programmatic financing approach that combined policy reform, institutional strengthening, and sectoral investments across multiple provinces. This layered approach mobilized public and private capital, strengthened environmental governance systems, and accelerated large-scale emission reductions.

Box: A Programmatic and Integrated Approach to Improving Air Quality in the Greater Beijing–Tianjin–Hebei Region, People's Republic of China

The Greater Beijing–Tianjin–Hebei (BTH) region comprises the municipalities of Beijing and Tianjin; Hebei, Henan, Liaoning, Shandong, and Shanxi provinces; and the Inner Mongolia Autonomous Region. At the time of program design, the region accounted for roughly one-third of the gross domestic product of the People's Republic of China (PRC) gross domestic product and was characterized by a dense population base and a high concentration of energy-intensive and polluting industries.

To address severe air pollution in this region, the Government of the PRC launched the Comprehensive Action Plan for Air Pollution Prevention and Control (2013–2017), followed by the Three-Year Action Plan for Winning the Blue Sky War (2018–2020). These plans outlined national priorities for air pollution prevention and control.

Implementation of the action plan required an estimated CNY249.03 billion in investment, far exceeding available domestic funding. Key constraints included insufficient public resources for pollution control investments, limited access to credit for small and medium-sized enterprises, and high upfront costs associated with cleaner and renewable technologies. Despite significant progress, structural challenges persisted, including weak enforcement, continued reliance on coal, and fragmented policy implementation across jurisdictions.

In 2015, the Government of the PRC and the Asian Development Bank (ADB) agreed to establish a multiyear, multisector lending program (2015–2020) to support implementation of these national plans. The program adopted a programmatic and integrated approach addressing policy reform, green financing, and technology deployment for both air quality improvement and greenhouse gas emissions reduction.

Between 2015 and 2020, ADB provided approximately $2.55 billion across eight loans to implement this approach. The program deployed a combination of financing modalities, including the following:

(i) A **policy-based loan** to support regulatory reforms and strengthen environmental governance in Hebei province.

(ii) Five **financial intermediation loans** to scale up pollution control investments, deploy advanced technologies in major emitting sectors, demonstrate green finance mechanisms, expand access to capital—particularly for small and medium-sized enterprises—and pilot financing structures that mobilize private, institutional, and commercial capital.

(iii) A **results-based loan** supporting a defined time period and geographic scope of the Cleaner Fuel Switch Investment Program issued by the Henan provincial government.

(iv) A **project loan** to finance advanced heating and cooling technologies in Shandong province, replacing coal with cleaner sources such as natural gas, renewable energy, and waste heat recovery.

continued on next page

[32] Chinese Academy of Environmental Planning and Nanjing University. 2015. *Air Pollution Prevention and Control Action Plan (2013–2017): Investment and Financing Needs and Impacts of Implementation.*

Box *continued*

This cross-sectoral approach strengthened the policy and regulatory framework for air quality management while establishing a **green financing platform** through the US$500 million *Air Quality Improvement in the Greater Beijing–Tianjin–Hebei Region: China National Investment and Guaranty Corporation's Green Financing Platform Project,* a financial intermediation loan under the program. The platform provided credit lines, sub-loans, and guarantees to support SMEs in transitioning to clean energy systems and to catalyze additional private investment.

Thirteen ADB **technical assistance** packages complemented the lending program, further enhancing institutional capacity and enabling long-term investment in clean air and low-carbon projects. The initiative also supported the issuance of the PRC's ***first domestic clean air bond in 2022***, aligned with the standards of the Climate Bonds Initiative and compliant with multilateral development bank safeguards requirements. The bond attracted significant interest from both international and domestic financial institutions and green bond issuers.

In parallel, ADB's Private Sector Operations Department approved nonsovereign financing linked to the broader initiative, including loans to Henan Tian Lun Gas Group and to Huaneng Tiancheng Financial Leasing, further mobilizing private capital for clean energy solutions.

Through coordinated action by central and local governments as well as development partners, **$PM_{2.5}$ levels in the region declined by 44.2%** in 2023 compared to 2015 levels. **Sulfur dioxide concentrations fell by 76.3%, and nitrogen oxide concentrations declined by 34.8%** over the same period. The proportion of days classified as having good air quality increased to 63.1%. The program also delivered significant climate co-benefits, avoiding an estimated **36.08 million tons of CO_2 equivalent** annually upon completion, relative to baseline scenarios.

Source: Zhang, L. and L. Zhou. 2025. Improving Air Quality in the People's Republic of China: Lessons from the Greater Beijing–Tianjin–Hebei Region. *ADB East Asia Working Paper Series.*

Moving forward, cities can benefit from a clearer understanding of their air quality investment needs and the financing mechanisms and instruments available to address them, tailored to their readiness, socioeconomic context, and the type of clean air intervention required. Financing options may range from grants and technical assistance for cities at early stages of air quality management capacity to concessional loans or blended finance for larger infrastructure investments.

6 City-Level Assessment and Priority Actions

This report synthesized the annual particulate matter ($PM_{2.5}$) trends, major emission sources, institutional capacity, and financing conditions to identify practical pathways to strengthen clean air action across Southeast Asia. The analysis highlights that cross-sectoral measures in transport, industry, agriculture, and waste management can accelerate progress in both air quality improvement and climate mitigation. The section below presents city-level summaries of the current air quality situation, management readiness, financing context, and priority interventions in eight urban centers within the Association of Southeast Asian Nations (ASEAN): Bangkok, Bogor, Ha Noi, Ho Chi Minh City, Iloilo City, Jakarta, Phnom Penh, and Quezon City. Detailed case studies are provided in Appendix 2.

BANGKOK, THAILAND	
Situation and Priority Sectors	Bangkok's annual $PM_{2.5}$ levels (2019–2024) consistently exceeded Thailand's national standard (15 μg/m^3) and all the World Health Organization (WHO) interim targets. Major emission sources include transport, industrial activity (including factories and power generation), and seasonal open agricultural burning, particularly during the dry season.
Management Capacity	Thailand's Pollution Control Department (PCD) has strengthened national policies and strategies addressing air quality and its climate co-benefits, while the Bangkok Metropolitan Administration (BMA) has implemented localized measures. These include enhanced vehicle emissions monitoring, promotion of clean transport, regulation of open burning, and deployment of digital tools like Air4Thai and BurnCheck for agricultural fire monitoring. Bangkok maintains a relatively comprehensive air quality management (AQM) network, jointly managed by PCD and BMA with more than 50 $PM_{2.5}$ monitoring stations.
Program and Initiative Financing Options	Public funding and technical support from ADB and international partners currently support major clean air initiatives. However, sustainable and long-term financing mechanisms remain limited, particularly for scaling clean technologies, incentive-based emissions reduction programs, and expanded monitoring systems. Thailand's National Sustainable Financing Framework provides a potential pathway but further localization and operationalization at the city level are needed.
Priority Actions	(1) **Transport sector.** Develop and implement an e-mobility roadmap, pilot e-*tuk-tuks*, expand electric vehicle (EV) charging infrastructure, and establish low-emission zones (LEZs). (2) **Agriculture sector.** Scale non-burning alternatives, strengthen community-based fire management, and enhance early warning systems to reduce seasonal haze.

BOGOR, INDONESIA	
Situation and Priority Sectors	The most recent $PM_{2.5}$ data for Bogor (2017) recorded 24-hour concentrations of 37 µg/m^3 at Bubulak Terminal. While this level met Indonesia's national standard at the time, it exceeded the WHO guideline levels. The absence of updated or long-term trend data underscores a significant gap in continuous air quality monitoring. The 2018 emissions inventory identified transport as the primary $PM_{2.5}$ source, followed by medical waste incineration and open burning. Industrial emissions remain comparatively limited due to the use of cleaner fuels.
Management Capacity	Bogor developed a clean air action plan (CAAP) in 2018 with support from academic and international partners, aligned with national air quality strategies. The city established a multi-stakeholder working group and collaborated with research institutions on emissions inventories and planning processes. However, implementation has been constrained by coordination challenges and limited regulatory enforcement capacity. Bogor operates a real-time AQM network with three fixed stations measuring $PM_{2.5}$, PM_{10}, and gases, with public data access. However, monitoring coverage is limited, and the city lacks mobile sensors, source apportionment studies, and emissions modeling capacity, restricting evidence-based policy decision and evaluation.
Program and Initiative Financing Options	Initial funding for air monitoring infrastructure was provided through public budgets and external technical support. Clean Air Asia through the Integrated Program for Better Air Quality Program supported the development of the Bogor CAAP 2018. Currently, Bogor does not have a dedicated or sustained clean air financing mechanism. Most activities rely on short-term project funding or academic partnerships, limiting the long-term sustainability and scale-up of interventions.
Priority Actions	CAAP 2018 sets forth a series of recommendations and potential measures across multiple sectors. These include: (1) **Transport emissions.** Enhance public transport systems and promote EV adoption. (2) **Agricultural emissions.** Promote non-burning alternatives for agricultural waste management. (3) **Waste management.** Pilot decentralized waste-to-energy solutions.

HA NOI, VIET NAM	
Situation and Priority Sectors	Ha Noi's annual average $PM_{2.5}$ levels (2021–2023) have consistently exceeded national standards and WHO AQG, indicating a need for strengthened interventions. This challenge is particularly pronounced during the winter season due to low atmospheric dispersion conditions. Ha Noi's AQMP 2024–2030 identified transport as the largest contributor (based on 2019 data) to $PM_{2.5}$ emissions, particularly from motorbikes, trucks, and taxis, followed by industry and agriculture.
Management Capacity	At the national level, the Ministry of Agricultural and Environment sets the overall policy direction, to which city-level actions must align. In Ha Noi, interagency coordination mechanisms are in place, particularly between the Department of Agriculture and Environment and the Department of Construction, especially on transport emissions management. Technical capacity remains constrained in key areas, including the preparation of emissions inventories, source apportionment studies, and the integration of air quality data into broader sustainable development planning. Ha Noi operates an established AQM network composed of fixed, mobile, and sensor stations, with active public dissemination of AQI data.
Program and Initiative Financing Options	Through partnerships with international organizations, Ha Noi has implemented a range of air quality management activities despite limited local budget allocations. Technical assistance, capacity building, policy analysis, and training programs have been supported by partners including Clean Air Asia.
Priority Actions	Ha Noi's AQMP 2024–2030, with a vision for 2035, outlines a multisectoral approach to reducing air pollution and improving public health. Key priority actions include: (1) **Transport emissions.** Expanding electric and green public transport, strengthening nonmotorized transport, and low-emission zones. (2) **Industry emissions.** Promoting cleaner production, particularly in craft villages. (3) **Agricultural emissions.** Piloting circular economy approaches for managing agricultural by-products. (4) **Waste management.** Developing waste treatment and waste-to-energy facilities.

HO CHI MINH CITY, VIET NAM	
Situation and Priority Sectors	Ho Chi Minh City's annual average $PM_{2.5}$ levels in recent years have remained above WHO AQG although the city met the national standard between 2020 and 2022. Studies identify transport as a primary source of $PM_{2.5}$ emissions along with industrial activities and domestic sources, particularly residential and commercial cooking.
Management Capacity	At the national level, MAE provides overall policy direction, to which city-level measures must align. Actions for the city included setting emissions standards for incinerators and improving motorbike inspection and control programs. Technical studies, such as emission inventories, have been developed in collaboration with local universities and experts due to the limited technical capacity of the local government. The city operates an AQM network composed of automatic and manual monitoring stations. Between 2021 and 2030, the city aims to expand continuous monitoring coverage, with a target of 20 stations.
Program and Initiative Financing Options	Partnerships established with the city have supported air quality management activities, including training and capacity building, and the development of the CAAP, delivered by Clean Air Asia. Other partners, such as ADB, United Nations Development Programme, World Health Organization, and C40 Cities, also provide ongoing support through air quality improvement projects.
Priority Actions	The CAAP, developed with support from ADB, identifies priority measures focused on the following: (1) **Industry emissions.** Improving enforcement of process emission standards and fugitive dust control and promoting alternative fuel use. (2) **Transport emissions.** Improving transport infrastructure and promoting the shift to EV motorcycles and alternative fuel use. (3) **Waste management.** Developing sustainable municipal waste strategies and waste-to-energy facilities. (4) **Agricultural emissions.** Improving management of agricultural waste and adopting alternative technologies.

ILOILO CITY, PHILIPPINES	
Situation and Priority Sectors	Iloilo City's recorded $PM_{2.5}$ levels in recent years generally remained within national standards but continued to exceed international health-based guidelines, including WHO AQG and interim targets. The limited AQM infrastructure in the city may not fully capture the extent of emissions. Available emissions data indicate that the transport sector is the primary source of urban air pollution, particularly due to emissions from older vehicles. Household fuel use also contributes to particulate pollution, along with emissions from waste burning, landfills, and road dust.
Management Capacity	Iloilo City has developed its Clean Air Plan (2015–2017) and is currently updating its Local Climate Change Action Plan (2025–2031) to reflect integrated strategies on air pollution and greenhouse gas (GHG) mitigation. Coordination among relevant departments and stakeholders supports the implementation of cross-sectoral actions. While the city has engaged in research collaborations and received training on mobility data governance, it still lacks continuous AQM stations and has limited capacity for emissions source apportionment and long-term data analysis. Integration of air quality data into policy and program development remains a gap.
Program and Initiative Financing Options	Iloilo City has demonstrated readiness for public–private partnerships in transport and waste projects. The city has also engaged regional and international partners for clean air and climate-related research. Financing remains largely dependent on external technical and donor support. In-house capacity for proposal development and emissions modeling is still limited.
Priority Actions	The city's draft LCCAP and clean air strategies emphasize the following: (1) **Transport emissions.** Improving public transport systems, promoting sustainable urban mobility, and supporting the transition to low-emission vehicles for government operations. (2) **Waste management.** Piloting decentralized waste-to-energy solutions.

JAKARTA, INDONESIA	
Situation and Priority Sectors	The air quality situation in Jakarta remains a significant concern in recent years, with consistently high $PM_{2.5}$ levels often exceeding both national standards and WHO AQG. Based on the available data from several stations in Jakarta from 2018 to 2024, the annual $PM_{2.5}$ concentrations ranged from 35.5 to 46 µg/m^3, with the highest values observed in 2018 (46.0 µg/m^3) and 2019 (45.7 µg/m^3). The 2025 research on the city's emissions profile indicates that road transport is the dominant source of $PM_{2.5}$ emissions, followed by the industry sector.
Management Capacity	Jakarta has established a range of policies and measures to curb emissions from transport and industry, with active stakeholder support. The Strategy for Air Pollution Control, introduced in 2022, includes measures focused on (i) strengthening air pollution control governance, (ii) reducing emissions from mobile sources, and (iii) reducing emissions from stationary sources. The AQM network in Jakarta is composed of 12 automatic monitoring stations that feed into the AQI reporting system.
Program and Initiative Financing Options	Technical aspects of air quality management in Jakarta have been supported by international partners and networks, partly in response to the city's air pollution challenges. External support has facilitated the development of monitoring systems, emissions inventories, and modeling tools, and has informed policy recommendations and capacity-building initiatives. Jakarta has also utilized climate-related financing opportunities including multilateral development bank support and international climate funds to advance clean transport, renewable energy, and emissions reduction initiatives.
Priority Actions	A set of recommended strategies for Indonesia, developed in May 2025 by Systemiq and the Bandung Institute of Technology, with support from Climate Works Foundation, outlines actions that can be initiated in Jakarta. These strategies focus on the following: (1) **Transport emissions.** Developing integrated public transport through route expansion, improved physical and service integration, and accelerating the adoption of EVs. (2) **Industry and energy emissions.** Upgrading or retrofitting the emission control systems of coal-fired power plants. (3) **Waste management.** Eliminating open burning and adopting long-term waste management solutions, including waste-to-energy initiatives.

PHNOM PENH, CAMBODIA	
Situation and Priority Sectors	Between 2017 and 2023, Phnom Penh recorded annual average $PM_{2.5}$ levels above WHO AQG, although concentrations remained within national standards. Based on studies of the national emissions profile, key sources of $PM_{2.5}$ emissions include residential, waste, charcoal production, transport, and industry sectors. Wood remains in use as a fuel source, particularly in the rural areas.
Management Capacity	The Ministry of Environment provides the overall policy direction and regulatory framework for air quality management, with city-level policies expected to align accordingly, including in the development of local sectoral plans. Implementing agencies continue to require capacity building. Multi-stakeholder engagement and support for air quality initiatives remain limited due to weak coordination mechanisms and the absence of sustained financing.
Program and Initiative Financing Options	At the national level, air quality management activities have been supported largely by international cooperation. This support has included technical training, capacity building, and the development of policies and action plans. Collaborative projects implemented under these arrangements include enhancements to the city's AQM network.
Priority Actions	Although Phnom Penh does not have a city-specific action plan, the Clean Air Plan for Cambodia, developed with support from Climate and Clean Air Coalition in 2021, provides guidance for city-level air quality interventions. Key actions include: (1) **Transport emissions.** Improving public transport strategy and route systems and promoting EV uptake for public transport. (2) **Agricultural emissions and waste management.** Promote non-burning alternatives for agricultural waste management. (3) **Industrial emissions.** Improving the efficiency of charcoal production.

QUEZON CITY, PHILIPPINES	
Situation and Priority Sectors	From 2022 to 2024, the annual average $PM_{2.5}$ concentrations in the city remained within national air quality standards but exceeded international health-based guidelines, including WHO AQG and interim targets. Available emissions data indicate that the transport sector is the main contributor to fine particulate pollution, with additional emissions coming from industrial operations, open burning, construction activities, and waste-related sources.
Management Capacity	The city has developed an AQMP aligned with broader climate strategies and international clean air commitments. It has established multisectoral coordination mechanisms to support implementation and decision-making. AQM capacity has improved through the installation of fixed monitoring stations and non-reference sensors, with data made publicly accessible and used in selected planning activities. However, challenges remain in fully integrating air quality data into policy and program development.
Program and Initiative Financing Options	Quezon City has mobilized both local budget allocations and donor-funded support for key programs, including AQMP development, sensor deployment, and pilot projects. Clean Air Asia supported the AQMP development in 2023. Long-term financing remains partly dependent on external partnerships, and institutionalized funding for air quality operations is still limited.
Priority Actions	The city's AQMP and climate plans outline sectoral actions. This includes: (1) **Transport emissions.** These cover implementing green zones, promoting public/mass transport, advancing fleet electrification, introducing vehicle management systems, and encouraging active school transport. (2) **Waste management.** This includes establishing a city-wide composting program and anaerobic biodigesters in public markets.

Path Ahead for ASEAN Cities

Across the eight ASEAN cities assessed, several consistent patterns emerge. Annual average $PM_{2.5}$ concentrations remain above the WHO guideline levels in most cities, even where national standards are met. Transport is the dominant source of urban particulate pollution across nearly all cases, followed by contributions from industry, open burning, agriculture, residential fuel use, and waste management. Seasonal factors, including dry-season agricultural burning and winter inversion conditions, intensify pollution episodes in several cities.

Institutional frameworks for air quality management are increasingly in place. Most cities have developed action plans or are in the process of updating them. Monitoring systems have expanded, and public disclosure of air quality data has improved. However, technical capacity remains uneven. Gaps persist in emissions inventory development, source apportionment analysis, emissions modeling, and integration of air quality data into urban and sectoral planning processes. In several cities, monitoring coverage remains limited, constraining data-driven decision-making.

Implementation gaps are also evident. While plans exist, interagency coordination across transport, energy, environment, agriculture, and urban development authorities remains a challenge. Regulatory enforcement varies in strength, and accountability mechanisms are not always clearly defined. As a result, many priority measures remain at early stages of implementation.

Financing represents one of the most significant cross-cutting constraints. Few cities have dedicated, sustained clean air financing mechanisms. Many interventions rely on short-term project-based funding, technical assistance, or donor support. Scaling priority measures such as transport electrification, cleaner production upgrades, waste treatment infrastructure, and alternatives to open burning will require predictable and long-term capital aligned with national development strategies and climate commitments. Strengthening project preparation capacity and developing bankable pipelines will be essential to mobilize diversified financing, including public budgets, concessional finance, private investment, and climate-related funding instruments.

At the same time, the report highlights a clear opportunity. Many priority interventions identified across the eight cities generate both air quality and climate benefits. Electrification of transport, cleaner industrial processes, improved waste systems, and reductions in open burning can simultaneously reduce $PM_{2.5}$ concentrations and greenhouse gas emissions. Aligning air quality strategies with broader climate, urban development, and sustainable finance frameworks can help unlock additional resources and enhance policy coherence.

Although improving air quality in ASEAN cities is a key environmental objective, it is also a public health priority, an economic imperative, and the foundation of resilient urban growth. The evidence presented in this report demonstrates that policy frameworks are emerging and technical foundations are being strengthened. The next phase will require accelerated implementation, stronger institutional coordination, improved data systems, and sustained financing at scale.

Appendix 1: Methodology

This report on *Air Quality in Southeast Asia: Status, Trends, and Priority Interventions Across Eight ASEAN Cities* provides information on air pollution trends, air quality management approaches in the region, and identifies priority actions. This appendix describes the data collection, quality assurance, and analytical methods applied in the report.

Scope and Limitations of the Report

The report analyzes government air quality data from 10 Association of Southeast Asian Nations (ASEAN) countries and more than 200 cities. Data were drawn from publicly accessible sources and from datasets provided by national ministries upon request. Official websites and other publicly available documents on sectoral policies and measures were also reviewed. A summary of the number of cities with available data per pollutant for each ASEAN country, based on any year with at least one valid observation, is presented in Table A1.1.

Table A1.1: Number of Cities with Available Air Quality Data Per Pollutant by ASEAN Country

Country	$PM_{2.5}$	PM_{10}	NO_2	SO_2	CO	O_3
Brunei Darussalam	2	4	2	2	2	2
Cambodia	25	0	10	1	0	10
Indonesia	26	24	19	20	17	18
Lao PDR	7	0	0	0	0	0
Malaysia	16	16	16	16	16	16
Myanmar	2	0	0	1	0	0
Philippines	57	77	19	23	13	19
Singapore	1	1	1	1	1	1
Thailand	81	69	46	37	1	4
Viet Nam	3	1	0	5	0	0

ASEAN = Association of Southeast Asian Nations, CO = carbon monoxide, Lao PDR = Lao People's Democratic Republic, NO_2 = nitrogen dioxide, O_3 = ground-level ozone, PM = particulate matter, SO_2 = sulfur dioxide.

Source: Clean Air Asia.

Based on Table A1.1, $PM_{2.5}$ exhibited the highest number of cities covered. Table A1.2 presents the names of cities with available $PM_{2.5}$ data, by country.

Table A1.2: Cities with Available $PM_{2.5}$ Data by ASEAN Country, 2011–2024

Country	Cities with Data
Brunei Darussalam	Belait, Brunei Muara
Cambodia	Banteay Meanchey, Battambang, Kampong Cham, Kampong Chhnang, Kampong Speu, Kampong Thom, Kampot, Kandal, Kep, Koh Kong, Kratié, Mondulkiri, Oddar Meanchey, Pailin, Phnom Penh, Preah Sihanouk, Preah Vihear, Prey Veng, Pursat, Ratanakiri, Siem Reap, Stung Treng, Svay Rieng, Takéo, and Tboung Khmum

continued on next page

Table A1.2 *continued*

Country	Cities with Data
Indonesia	Aceh, Balikpapan, Banjarmasin, Batam, Bogor, Central Jakarta, Denpasar, Gorontalo, Jakarta, Jambi, Kab. Bekasi, Kab. Bogor, Kab. Kotawaringin Timur, Kab. Tangerang, Kendari, Makassar, Manado, Mataram, Padang, Palangkaraya, Palembang, Pekanbaru, Pontianak, South Jakarta, and Tangerang Selatan
Lao PDR	Bokeo, Khammouane, Luang Namtha, Luang Prabang, Oudomxay, Savannakkhet, Vientiane
Malaysia	Johor (state), Kedah (state), Kelantan (state), Melaka (state), Negeri Sembilan (state), Pahang (state), Perak (state), Perlis (state), Pulau Pinang (state), Sabah (state), Sarawak (state), Selangor (state), Terengganu (state), Wilayah Persekutuan Kuala Lumpur, Wilayah Persekutuan Labuan, and Wilayah Persekutuan Putrajaya
Myanmar	Mandalay and Yangon
Philippines	Antipolo, Bacolod, Baguio, Balanga, Batac, Batangas, Biñan, Butuan, Cabadbaran, Cagayan de Oro, Caloocan, Dagupan, Davao, El Salvador, General Santos, Iligan, Iloilo City, Isabel, Koronadal, La Trinidad, La Union, Laganes, Laguna, Las Piñas, Legazpi, Lingayen, Makati, Malabon, Manila, Marikina, Metro Manila, Meycauayan, Sto. Tomas, Muntinlupa, Naga, Navotas, Olongapo, Ormoc, Parañaque, Pasay, Pasig, Pateros, Pavia, Puerto Princesa, Quezon, San Carlos, San Fernando, San Juan, Santa Rosa, Tacloban, Tagoloan, Taguig, Urdaneta, Valenzuela, Vigan, Villanueva, and Zamboanga
Singapore	Singapore
Thailand	Amnat Charoen, Aranyaprathet, Ayutthaya, Bang Sao Thong, Bangkok, Betong, Bo Win, Bueng Kan, Buriram, Chaiyaphum, Chang Phueak, Chanthaburi, Chiang Mai, Chiang Rai, Chonburi, Hat Yai, Hua Hin, Huai Kon, Kalasin, Kamphaeng Phet, Kanchanaburi, Khlong Luang, Khon Kaen, Laem Chabang, Lampang, Lamphun, Loei, Lop Buri, Mae Hong Son, Mae Mo, Mae Sai, Mae Sot, Map Ta Phut, Mukdahan, Nai Mueang, Nakhon Nayok, Nakhon Pathom, Nakhon Phanom, Nakhon Ratchasima, Nakhon Sawan, Nakhon Si Thammarat, Nan, Narathiwat, Nong Khai, Nonthaburi, Om Noi, Pak Kret, Pak Nam, Phayao, Phetchabun, Phichit, Phitsanulok, Phra Pradaeng, Phrae, Phuket, Plaeng Yao, Pluak Daeng, Ratchaburi, Rayong, Sakon Nakhon, Samut Prakan, Samut Sakhon, Samut Songkhram, Saraburi, Satun, Si Phum, Sing Buri, Sisaket, Sri Maha Pho, Sukhothai, Suphan Buri, Surat Thani, Surin, Trang, Trat, Ubon Ratchathani, Udon Thani, Uthai Thani, Uttaradit, Yala, and Yasothon
Viet Nam	Ha Noi, Ho Chi Minh City, and Hoa Binh

ASEAN = Association of Southeast Asian Nations, PM = particulate matter.
Source: Clean Air Asia.

The research and analysis conducted for this report have the following limitations:

a. **Variation in publicly available data across different platforms.** The specific type and extent of air quality data shared publicly vary by country and platform. The pollutants reported per monitoring site differ, data formats are not uniform, and historical coverage is inconsistent. For instance, Singapore has 22 fixed monitoring sites, but only the national average air quality data are publicly accessible. In Indonesia, only real-time data are available through the official government website, with no access to historical records. Some government websites allow direct download of data spreadsheets, while others provide information only in PDF or figure format, particularly if sourced from national statistics or environmental reports.

b. **Data not reported in pollutant concentrations.** In some countries, air quality data are not reported in terms of pollutant concentrations. For example, Cambodia reports air quality data using the air quality index (AQI), while Brunei Darussalam reports using the pollutant standard index (PSI). As this report analyzes pollutant concentrations, AQI and PSI based data were excluded.

c. **Language constraints in data collection.** The data collection process was limited to information that was accessible and translatable to the English language. Government reports and datasets only available in the local language and not readily translatable were not included in the analysis.

d. **Variations in geographic scope and pollutant reporting.** In some countries, air quality data are aggregated at the state or country level rather than reported by individual monitoring sites. For analyses requiring city-level data, state or country averages were used as proxies. However, locations that could not be spatially validated (i.e., monitoring sites that were not clearly labeled or described) were excluded from map generation.

Data Availability and Accessibility

This report refers to "available" air quality data if the information exists, and "accessible" air quality data when the information can be readily obtained by the public. Data coverage varies by country depending on what government reports (e.g., reporting of the number of monitoring sites managed versus the public availability of detailed air quality data from individual monitoring sites). For purposes of this report, the analysis focused on accessible air quality data.

Data Collection

The data used in the report comes from official government reports, websites, and statistics, supplemented with other reputable reports and journal publications, as well as data and information directly shared by government agencies upon official requests. The report integrated information from the Clean Air Asia air quality database, compiled from 2011 to 2024 and updated as of January 2026. Key air quality parameters such as particulate matter ($PM_{2.5}$ and PM_{10}), sulfur dioxide (SO_2), nitrogen dioxide (NO_2), carbon monoxide (CO), and ground-level ozone (O_3) annual concentrations were compiled and organized by pollutant, monitoring site, city/state/province, and country, depending on the availability of site-specific data. In addition to AQM data, policies and plans related to air quality management were collected, with a focus on ambient air quality standards and guidelines, emission inventories, health impact-related action plans, vehicle and emission standards, and CAAPs. The geographic scope of the data collected encompasses 10 ASEAN countries, with varying levels of detail depending on data availability.

Table A1.3: Data Collected for the Report

Data	Description
Country	Country where the monitoring site is located.
City	City where the monitoring site is located; can also be the province, county, or state depending on availability of data.
Monitoring Site Name	Monitoring station name or site ID.
Pollutant Concentration	Measured concentration of air pollutants ($PM_{2.5}$, PM_{10}, SO_2, NO_2, CO, and O_3) in micrograms per cubic meter ($\mu g/m^3$), milligrams per cubic meter (mg/m^3), parts per million (ppm), or parts per billion (ppb).
Data Source	Link to the government report or website used as reference for the data.
Air Quality Standards	National air quality standards and/or guideline values per pollutant.
Air Quality Management Legal Framework	Information on Clean Air Act, National and Subnational Action Plan, and National and Subnational Programs for Air Quality.
Government Air Quality Database and National Report	Link of the website or the report, if available/accessible.
Government Air Quality Sites	Number of monitoring sites per country.

continued on next page

Table A1.3 *continued*

Data	Description
Population	Population data in countries and megacities, as per the United Nations Population Division.
Air Pollution Sources	Information from international projects and programs, academic publications, national emission inventories, and published source apportionment studies.
Air Quality and Health Policies	Country policies and action plans related to air pollution and health.
Health Impact Attributed to Air Pollution	$PM_{2.5}$-attributed mortality per country, as per the 2021 Global Burden of Disease Report.
Country Income	Country income (high, upper/lower middle, low), as per the World Bank.
Policies and Measures	• Country regulations on fuel standards and vehicle emission standards, and initiatives for adopting electric vehicles. • Country regulations on industrial emissions standards and policies for transitioning to cleaner fuel alternatives and renewable sources of energy. • Sustainable agricultural practices. • Country regulations for municipal waste open burning.

CO = carbon monoxide, NO_2 = nitrogen dioxide, O_3 = ground-level ozone, PM = particulate matter, SO_2 = sulfur dioxide.
Source: Clean Air Asia.

Data Processing and Quality Assurance

In the analysis of air quality data, only outdoor (ambient) monitoring stations with at least 75% data coverage were included. This threshold represents the fraction of the year for which pollution concentration data were reported, i.e. "annual" data must have a minimum coverage of 275 out of 365 days. Ensuring dataset completeness is essential to determine whether an average value is representative of the period being measured.

Annual arithmetic means were calculated for stations that reported only daily or hourly averages. Zero values, negative values, and entries flagged or evaluated as invalid by the data provider were excluded from the calculations. All data reported by government agencies were assumed to be from regularly calibrated reference air quality instruments. Pollutant concentrations were expressed in µg/m^3 for $PM_{2.5}$, PM_{10}, SO_2, NO_2, and O_3, and mg/m^3 for CO. Unit conversions were performed using the conversion factors provided in the UK Air Information Resources of the UK Department for Environment, Food & Rural Affairs to ensure uniformity and comparability. In cases where only nitrogen oxides (NO_x) values were reported, these were assumed to be equivalent to NO_2 for comparison with WHO AQG. All pollutant measurements were assumed to be under standard conditions (25°C and 1013 mbar or 1 atm). Pollutant concentrations were analyzed by monitoring site and country for each year with available data. Analysis and plots were done using Microsoft Excel and R, while QGIS was used to map pollution data for monitoring sites with available geospatial information.

As this report applies a quantitative approach in analyzing air pollution concentrations, air quality data presented as indices, such as AQI, PSI, and air pollution index, were excluded to ensure comparability across cities and countries.

Appendix 2: Case Studies

Bangkok

Air Quality Status and Trends

Bangkok's most recent city-level air quality data covers 2019 to 2024.[1] For all recorded years, the city's annual average $PM_{2.5}$ concentrations exceeded both the World Health Organization (WHO) Air Quality Guidelines (AQG) (5 µg/m^3) and all interim targets (Its) (IT-1: 35 µg/m^3, IT-2: 25 µg/m^3, IT-3: 15 µg/m^3, and IT-4: 10 µg/m^3). Particulate matter ($PM_{2.5}$) levels also consistently exceeded Thailand's national annual standard of 15 µg/m^3 in the tallied years (Figure A2.1).[2]

Figure A2.1: Bangkok's Annual Average $PM_{2.5}$ Levels, 2019–2024

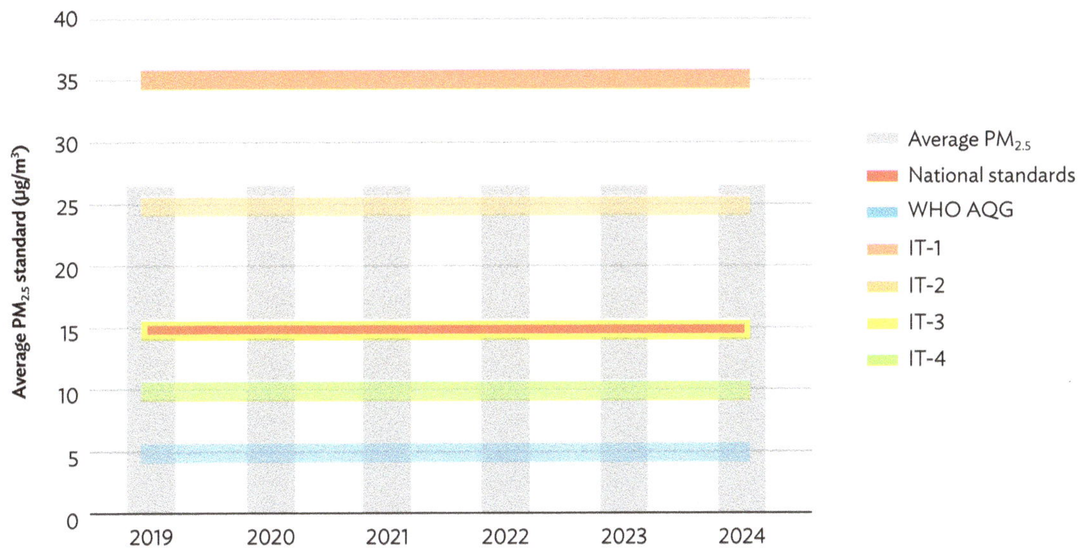

AQG = Air Quality Guidelines, IT = interim target, PM = particulate matter, WHO = World Health Organization.
Source: Clean Air Asia.

Studies assessing Bangkok's emissions profile have consistently identified transport, industrial, and agricultural sectors as the top contributors of air pollution. Based on the city's 2015[3] and 2019[4] emissions inventory (Figure A2.2), old in-use vehicles, factories, and power plants are the top emitters of $PM_{2.5}$. The seasonal agricultural waste burning, which typically occurs from November to April, further elevates $PM_{2.5}$ during the dry season.

1 Clean Air Asia's Compilation of Data for ASEAN.
2 *The Nation. 2022.* Thailand Adopts $PM_{2.5}$ Safety Standard on Par with Developed Nations. News release. 14 July.
3 Nguyen, T. K. O. 2017. *A Study in Urban Air Pollution Improvement in Asia*.
4 Asian and Pacific Centre for Transfer of Technology. 2024. *Technological Interventions and Gaps in Air Pollution Control in Bangkok, Thailand (Report 1)*.

Figure A2.2: Bangkok's Emissions Inventory on $PM_{2.5}$ in 2015 vs. 2019

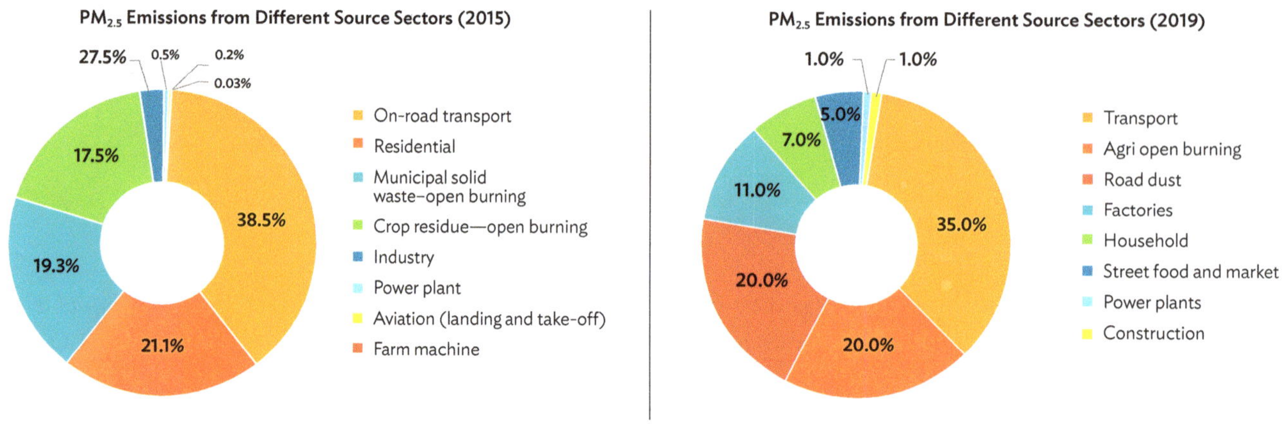

PM = particulate matter.
Source: Clean Air Asia.

Table A2.1: Bangkok's Drivers and Barriers to Clean Air

	Drivers	Barriers
Policy and Regulatory Framework	• A range of local policies and measures addressing emissions control and public health risks from air pollution.[a] • Alignment between national and local air quality action plans, with $PM_{2.5}$ recognized as a national priority.[b]	• Absence of comprehensive national legislation on air quality management (clean air bill pending approval). • Policy gaps, including the lack of vehicle age limits and incentives to promote newer, efficient vehicles and alternative agricultural waste management practices.
Institutional Capacity and Coordination	• Established committee that oversees monitoring, analysis, coordination, and public health communication ($PM_{2.5}$ Air Pollution Prevention and Solution Committee and the $PM_{2.5}$ Air Pollution Coordination and Resolution Center). • Active multistakeholder participation in city air quality and climate project planning and implementation.	• Coordination challenges between national and local agencies.
Data and Technology	• Comprehensive air quality monitoring (AQMt) system with real-time reporting. • Pilot projects and technologies implemented, including BurnCheck for agricultural waste burning.	• Lack of an integrated data sharing system among provinces for controlling agricultural open burning.
Financing and Resource Mobilization	• National Sustainable Financing Framework in place.[c]	• Limited sustainable finance, including the need for a dedicated budget to support incentives and deployment of clean tech.

PM = particulate matter.
[a] PR Bangkok. 2025. PR-Bangkok.
[b] Asian and Pacific Centre for Transfer of Technology. 2024. *Technological Interventions and Gaps in Air Pollution Control in Bangkok, Thailand (Report 1)*. United Nations Economic and Social Commission for Asia and the Pacific.
[c] Bangkok Metropolitan Administration. 2020. *City of Bangkok Sustainable Financing Framework*.
Source: Clean Air Asia.

Table A2.2: Bangkok's City Vision on Clean Air and Climate Commitments

Clean Air and Climate Commitments	Targets
Breathe Cities Initiative.[a] Bangkok will receive support in gathering air quality data, building capacities in air quality management, and enhancing community to advance existing clean air initiatives.	Reduce air pollution by up to 30% by 2030.
Indore 3R Declaration of Asian Mayors on Achieving Clean Water, Clean Land and Clean Air in Cities.[b] Bangkok signed a binding agreement between agencies at the city and urban government level to employ the 3R principle in resource management.	

3R = reduce, reuse, recycle.

a Clean Air Fund. 2024. Bangkok Breathe Cities.

b Government of Japan, Ministry of the Environment. 2015. *Results of the Third 3R Policy Phase: Thailand [PDF]*. and *The Nation*. 2023. Indore Mayors' 3R Declaration: Bangkok Signs Binding Urban Resource Agreement. News release. 28 February.

Source: Clean Air Asia.

Table A2.3: Bangkok's Proposed Air Quality and Climate Action

Priority Issues/Sector	Scope of Activities	Objective/Targets	Time Frame	Stakeholders
Transport Emissions	**Develop and implement e-mobility road map.** • Identify e-bus corridors and electric vehicle (EV) infrastructure needs. • Pilot e-*tuk-tuk* deployment. • Install charging stations in public transport hubs.	• Increase the share of EVs in public and private transport. • Reduce transport emissions.	>5 years	• Bangkok Metropolitan Administration • Bangkok Mass Transit Authority
	Implement low-emission zone (LEZ) and nonmotorized transport (NMT) systems, such as public bike rental services.	• Pilot LEZ sites implemented. • Reduce transport emissions.	1–5 years	• Bangkok Metropolitan Administration
Agricultural Emissions	**Promote alternative technologies and practices for agricultural waste management.** • Strengthen community-based fire management and early warning systems. • Conduct capacity-building activities.	• Pilot sites for alternative technologies established. • Reduce $PM_{2.5}$ emissions from agriculture, particularly during the dry season.	1–5 years	• Bangkok Metropolitan Administration • Provincial Government • Department of Forestry • Department of Agriculture • Pollution Control Department

PM = particulate matter.

Source: Clean Air Asia.

Bogor

City Air Quality Status and Emissions Profile

Due to limited historical data, 2017 remains the only year with publicly available $PM_{2.5}$ and PM_{10} measurements for Bogor. Monitoring was conducted across six locations, with the highest $PM_{2.5}$ concentration recorded at Bubulak Terminal, reaching 37 $\mu g/m^3$.[5] While this value remains below the national limit, it exceeds WHO's 2021 guideline of 15 $\mu g/m^3$ for 24-hour $PM_{2.5}$ exposure. Based on Bogor's 2016 greenhouse gas (GHG) inventory, the transport sector is the primary contributor to $PM_{2.5}$ emissions, largely driven by fossil fuel combustion. Other sources include medical waste incineration in hospitals, accounting for approximately 6%, and the open burning of garbage and post-harvest straw, which contributes around 5%. Industrial sources make up only 2% of emissions, as most facilities rely on natural gas, which produces relatively low particulate emissions.

Figure A2.3: Bogor's $PM_{2.5}$ Emissions Distribution per Sector, 2016

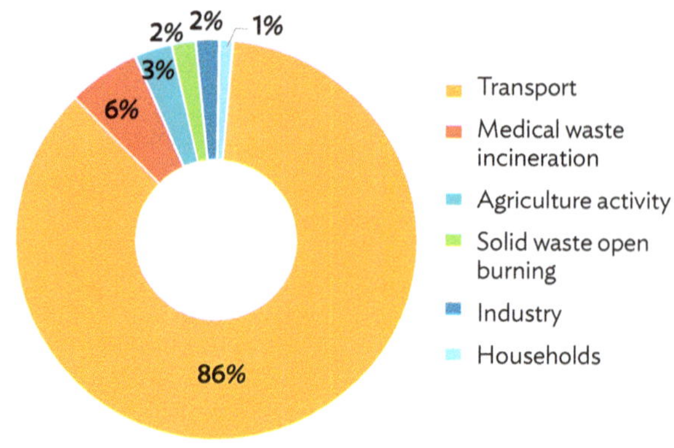

PM = particulate matter.
Source: Clean Air Asia.

Table A2.4: Bogor's Drivers and Barriers to Clean Air

	Drivers	Barriers
Policy and Regulatory Framework	• Provincial-level laws and regulations that support air quality management in the city. • Wide range of local regulations addressing climate change and air pollution, including the promotion of low-carbon emissions and innovation.[a] • Local clean air action plan targeting major pollution sources developed with stakeholder support	• Limited technical capacity and funding at the local level to operationalize plans. • Absence of a dedicated clean air bill at the national or regional level.
Institutional Capacity and Coordination Mechanism	• Capacity building supported through the Integrated Program for Better Air Quality program including clean air action plan (CAAP) development. • Active multistakeholder engagement in air quality management planning and CAAP development.	• Limited technical capacity within local institutions. • Weak implementation and interagency enforcement.
Data and Technology	• Pilot projects implementing sustainable waste management technologies. • Air quality reporting available through government service channels.	• Limited AQM capacity with coverage restricted to three monitoring stations and low spatial resolution.
Financing and Resource Mobilization	• Sustainable low-emission transport projects prioritized with budget approval from city council. • Funding options available through regional loans, municipal bonds, and public–private partnerships.	• Limited engagement with private sector partners.

AQM = air quality management.
[a] Circle Economy Foundation. 2022. Urban Low Emission Development Strategies–Bogor.
Source: Clean Air Asia.

[5] A.D. Permadi et al. 2018. *Clean Air Action Plan: Bogor City*. Institut Teknologi Nasional Bandung & Institut Pertanian Bogor.

Table A2.5: Bogor's City Vision for Clean Air and Climate Commitments

Clean Air and Climate Commitments	Targets
Integrated Program for Better Air Quality Program. This aims to strengthen technological and institutional capacity in air quality management and climate mitigation. It is supported by Japan's Ministry of Environment. Through this initiative, Bogor's clean air action plan (CAAP) is aligned with its midterm national development plan.	Reduce emissions from the industrial and transport sectors; improve air quality management systems and data management.
BreatheLife Campaign.[a] Bogor joined the BreatheLife network in 2019 following the development of the city's CAAP with support from the IBAQ Program.	Meet the WHO air quality targets by 2030.
Urban-LEDS II.[b] Bogor committed to implementing Low Emission Development Strategies from 2016 to 2021, with a focus on emerging cities in least developed countries.	Develop an integrated climate action plan to reduce greenhouse gas emissions.

[a] BreatheLife. 2019. Bogor City Develops Clean Air Action Plan, Intends to Join BreatheLife Network.
[b] Urban LEDS. 2019. Bogor City Renews Commitment to Implement Urban LEDS II.
Source: Clean Air Asia.

Table A2.6: Bogor's Proposed Air Quality and Climate Actions

Priority Issues/Sector	Scope of Activities	Objective/Targets	Time Frame	Stakeholders
Transport Emissions	**As part of the national electrification target, Bogor is included among the 11 priority cities to adopt 6,600 units of e-buses to reduce greenhouse gas.**[a] Angkot conversion through "Buy the Service" scheme.	• Reduce greenhouse gas emissions by 25% compared to business-as-usual (BAU) conditions.	>5 years	• Ministry of Transportation • Bogor City Department of Transportation • Biskita Operators • Angkot Operators/ Owners
	Construct the Bogor tram integrated with the Greater Jakarta LRT: project initiated in 2024 but construction timeline not yet defined.[b]	• Accelerate sustainable urban transport development and improve public transport systems	Ongoing (no timeline identified)	• Ministry of Transportation • Bogor City Department of Transportation • Institute for Transportation and Development Policy Indonesia
Agricultural Emissions	**Implement zero burning practices in rice field management.** • Formulate policies to ban open burning. • Conduct research on the use of straw as alternative fuel in manufacturing and construction. • Provide training to farmers.	• Reduction in GHG and air pollutants.	1 year	• Bogor City Agriculture Office • Bogor Environmental Management Agency

continued on next page

Table A2.6 *continued*

Priority Issues/Sector	Scope of Activities	Objective/Targets	Time Frame	Stakeholders
Waste Management	**Implement a waste-to-energy program.** Pilot community-scale waste management using hydro-pyrolysis technology.[c]	• Scale up pilot projects.	Proposed	• Bogor Environmental Management Agency

[a] Institute for Transportation and Development Policy, Indonesia. 2024. ITDP Indonesia Launches National Roadmap and Incentive Program for Road Based Public Transport Electrification.
[b] Nizar, A. 2024. Bogor to Develop Tram as Public Transportation Integrated with Greater Jakarta LRT. *SEA Today*. News release. 17 January.
[c] ICLEI Southeast Asia. 2021. Bogor City Puts Up Community-Scale Waste Management Pilot Project Using Hydro-Pyrolysis Technology.
Source: Clean Air Asia.

Ha Noi

City Air Quality Status and Emissions Profile

Based on the Ha Noi AQMP, annual average $PM_{2.5}$ concentrations from 2018 to 2020 nearly doubled the allowable amount (25 μg/m^3) as specified by the national technical regulation. Meanwhile, based on Clean Air Asia's database, $PM_{2.5}$ levels over the period of 2016 to 2022 (except for 2019) have consistently exceeded the national limits and WHO IT-1 of 35 μg/m^3.

Figure A2.4: Ha Noi's $PM_{2.5}$ Levels Against Various Air Quality Standards, 2021–2023

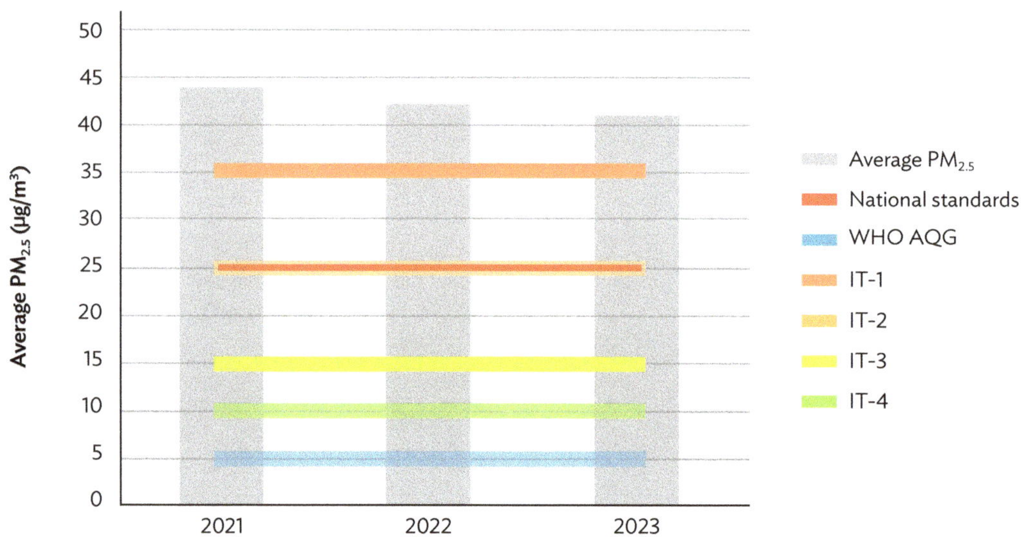

AQG = Air Quality Guidelines, IT = interim target, PM = particulate matter, WHO = World Health Organization.
Source: Clean Air Asia.

Ha Noi's AQMP identifies road vehicles (particularly motorbikes, trucks, and taxis), industry, and straw burning as the main contributors to $PM_{2.5}$ in the 2019 base year.

Table A2.7: Ha Noi's Drivers and Barriers to Clean Air

	Drivers	Barriers
Policy and Regulatory Framework	• National frameworks and local action plans aligned across key emission sectors, i.e., transport, agriculture, energy, and waste. • Proactive adoption of the national road map for vehicle emission standards.	• Limited sector-specific targets to guide the design and implementation of emission reduction measures. • Limited integration of air quality with climate change and sustainable development goals. • Lack of clear enforcement mechanism for vehicle emission standards, including coordinated monitoring systems across cities and provinces.[a] • Weak enforcement and limited application of sanctions or penalties for handicraft and industrial establishment.
Institutional Capacity and Coordination	• Existing interagency coordination for air quality management (AQM), i.e., Department of Agriculture and Environment and Department of Transport	• Absence of interregional coordination mechanism. • Limited technical capacity at all levels of administration, including within inner-city communities. • No evidence of multistakeholder engagement in air quality and climate policy development. • Inadequate infrastructure in suburban districts, constraining waste collection and transport
Data and Technology	• Established AQM network with Air Quality Index (AQI) reporting. • Accessibility of innovative technologies and alternative practices to reduce emissions in selected sectors (e.g., household emissions,[b] agricultural emissions).[c]	• Reliance on independent research or external partners to develop emissions inventories; a budget proposal was made for emission inventory development in 2025. • High investment cost for timely upgrading and maintenance of AQM systems. • High costs of advanced technologies for small-scale agricultural production.[d]
Financing and Resource Mobilization	• State budget allocations for priority programs and projects. • Engagement with international organizations and research institutions on selected air quality management initiatives.	• Limited in-depth cost–benefit analysis of emission reduction measures for decision-making and prioritization. • Public opposition to zoning fees for high-emission vehicles.

[a] *Vietnam News*. 2025. Emissions Regulations Present Challenges for State Agencies and Businesses. News release. 17 June.
[b] C40 Cities. 2018. Hanoi–Households Emissions Reduction Through Cookstove Conversions (Case study).
[c] Clean Air Asia and UNEP. 2021. Open Burning Case Study in Vietnam (Case study).
[d] Anh Kiet. 2024. Hanoi Applies High-Tech in Urban Agricultural Development. *Hanoi Times*. 4 December.

Source: Clean Air Asia.

Table A2.8: Ha Noi's City Vision for Clean Air and Climate Commitments

Clean Air and Climate Commitments	Targets
Ha Noi Air Quality Management Plan 2024–2030, with a Vision for 2035. Identifies 33 short- to medium-term priority programs and projects across key sectors (transport, agriculture, energy, and waste). It includes a road map for reviewing targets, assessing implementation effectiveness, and updating the plan.	• Maintain good and average levels of VN-AQI at least 75%–80% of the year. • Reduce $PM_{2.5}$ emissions by 20% compared to 2019 levels and manage $PM_{2.5}$ levels at <40 µg/Nm³ in urban areas, and <35 µg/Nm³ in suburban areas by 2030.

Nm_3 = normal cubic meter, PM = particulate matter, VN-AQI = Viet Nam Air Quality Index.
Source: Clean Air Asia.

Table A2.9: Ha Noi's Proposed Air Quality and Climate Action

Priority Issues/Sector	Scope of Activities	Objective/Targets	Time Frame	Stakeholders
Transport Emissions	**Promote the shift to EV and install additional EV charging stations within the city.**	• Achieve 50% share of EVs for bus fleet.	>5 years	• Department of Construction
	Implement LEZs in select districts of Ha Noi and during high pollution episodes. Pilot low-emission pedestrian zones.	• A pilot LEZ is being considered for selected areas. • Expand pedestrian zones to gradually reduce motorized vehicles and enhance cultural areas.	Ongoing 1 year	• Department of Agriculture and Environment
	Promote public transport use. • Expand the Ha Noi metro system with inclusive design infrastructure. • Introduce minibus routes.	• Ha Noi aims to develop 97 km of metro lines (24% progress in the master plan) from 2024 to 2030.	Ongoing >5 years	• Department of Construction
Agricultural Emissions	**Develop pilot projects using circular economy approaches to manage agricultural by-products.** Evaluate the efficiency of biogas plants in livestock farms to propose management solutions (implemented under the Bioenergy for Circular Agriculture [BeCA] project during 2021–2024).[a]	• Scale up biogas-to-electricity generation at medium and large livestock farms.	Ongoing >5 years	• Department of Agriculture and Environment
Waste Management	**Develop waste treatment and waste-to-energy facilities.** • Expand treatment facilities in Soc Son and Son Tay. • Develop methane capture systems for landfills.[b]	• Establish 17 waste treatment zones under the city's master plan for 2030 (vision to 2050). *(implementation delays reported due to site clearance issues.)*	Ongoing >5 years	• Department of Agriculture and Environment • Department of Construction

continued on next page

Table A2.9 *continued*

Priority Issues/Sector	Scope of Activities	Objective/Targets	Time Frame	Stakeholders
Industry and Energy Emissions	**Reduce emissions from craft villages.** • Switch to cleaner fuels. • Reduce coal and biomass use in boilers and furnaces. • Strengthen inspection and monitoring efforts to ensure compliance in handicraft villages.	• Reduce 524 tons of $PM_{2.5}$ emissions from craft villages.	Planned	• Department of Industry and Trade
	Promote cleaner production programs in the industrial sector. Continue the implementation of the National Action Program on sustainable production and consumption.	• By 2030, 40%–50% of localities will have plans to convert existing industrial parks to eco-industrial parks, and up to 10% of localities will plan new eco-industrial parks to attract investment.	Planned >5 years	• Department of Industry and Trade

EV = electric vehicle, km = kilometer, LEZs= low-emission zones.

[a] Funded by the Australian Government through the Business Partnerships Platform; also supported by HD Bank.

[b] World Bank. 2025. *Clean Air for Hanoi: What Will It Take?*

Source: Clean Air Asia.

Ho Chi Minh City

City Air Quality Status and Emissions Profile

Between 2016 and 2019, Ho Chi Minh City's $PM_{2.5}$ levels have exceeded Viet Nam's national standards for annual average concentration levels.[6] From 2019 to 2022, the city maintained its $PM_{2.5}$ levels just below the national threshold. However, $PM_{2.5}$ levels remained above the WHO guideline values, indicating continued public health risks and the need for stronger air quality management measures.

Figure A2.5: Ho Chi Minh City's Annual Average $PM_{2.5}$ Levels, 2016–2022

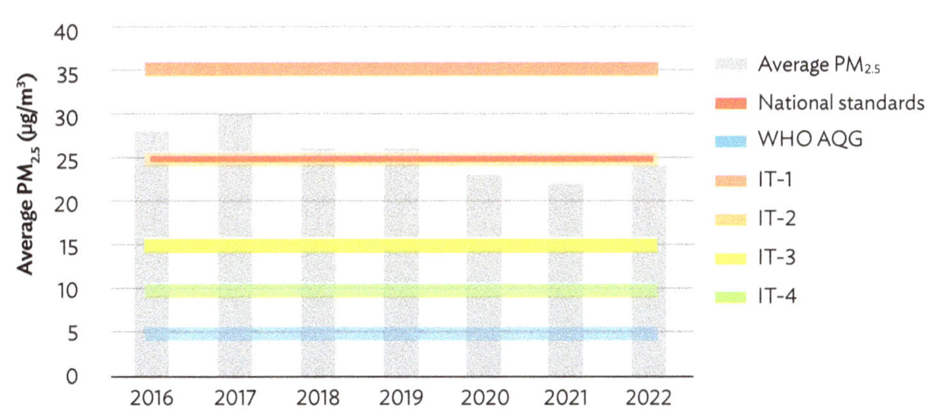

AQG = Air Quality Guidelines, IT = interim target, PM = particulate matter, WHO = World Health Organization.
Source: Clean Air Asia.

[6] Clean Air Asia. 2025. *2025 Air Quality in Southeast Asia: Status and Trends.*

Although no official emissions inventory has been developed for city planning purposes, findings of this report indicate that transport is the primary source of $PM_{2.5}$ emissions in the city, with additional contributions from industrial activities and residential and commercial cooking. Figure A2.6 presents a breakdown of $PM_{2.5}$ emissions by type of vehicle, showing that trucks are the largest contributors.[7]

Figure A2.6: Distribution of Ho Chi Minh's $PM_{2.5}$ Emissions from Vehicles

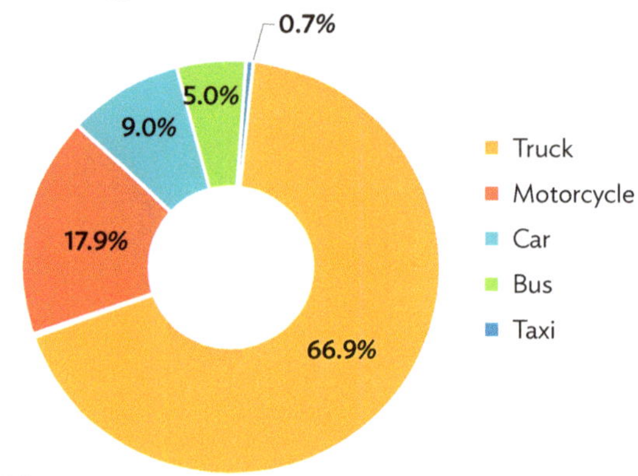

PM = particulate matter.
Source: Clean Air Asia.

Table A2.10: Ho Chi Minh City's Drivers and Barriers to Clean Air

	Drivers	Barriers
Policy and Regulatory Framework	• Established legal frameworks for air pollution control and air quality management. • National and city-level action plans aligned on climate and air quality objectives (National Plan on Air Quality Management 2021–2025). • City-level initiatives in support of national policies and regulations, including vehicle emissions inspection (motorbike inspection pilot program). • City-level standards for incinerator emissions (using nonhazardous industrial wastes as fuel).	• Gaps in the enforcement and monitoring of emission standards and vehicle inspection programs.
Institutional Capacity and Coordination	• Dedicated department for environmental management with subdepartments involved in various air quality management activities (previously under the Department of Natural Resources and Environment).	• Limited technical capacity and human resources, including for the enforcement of environmental regulations. • Limited stakeholder involvement in air quality control actions.
Data and Technology	• Deployment of low-cost sensors to complement the city's ambient AQM network, in collaboration with research institutions.[a]	• AQM data and emissions inventories are not publicly available. • No long-term or institutionalized funding to sustain AQM activities (i.e., air quality sensors and technical experts).

continued on next page

[7] Babela, D. et al. 2023. Clean Air Action Plan for Ho Chi Minh City. Ricardo.

Table A2.10 *continued*

	Drivers	Barriers
Financing and Resource Mobilization	• Opportunities for corporate social responsibility funding and private sector investment.[b] • Access to project financing through established international partnerships (e.g., UNDP and WHO,[c] Clean Air Asia, C40 Cities).	• Existing funds primarily allocated for inspection and monitoring programs.

AQM = air quality management, UNDP = United Nations Development Programme, WHO = World Health Organization.

[a] MIT University. Air quality monitoring stations in Ho Chi Minh City (Vietnam). In Sustainability–Sustainable Development Goals.

[b] From the "Ho Chi Minh City Key Air Quality Programs," presentation by Ngo Nguyen Ngoc Thanh in Bangkok in September 2024.

[c] Ministry of Agriculture and Environment, Government of Viet Nam. 2024. UNDP–WHO Supports Viet Nam in Addressing Air Pollution.

Source: Clean Air Asia.

Table A2.11: Ho Chi Minh's City Vision for Clean Air and Climate Commitments

Clean Air and Climate Commitments	Targets
Clean air action plan for Ho Chi Minh City (HCMC CAAP) for 2022–2030. The CAAP aims for the following: "The City of Ho Chi Minh City will work to reduce its pollutant emissions by focusing on implementing solutions, tasks, and priority projects, in order to raise awareness of, and improve, air quality." The plan includes sectoral road maps for transport, industry, and construction, as well as broader air quality management, aligned with the objectives of the National Plan on Air Quality Management, 2021–2025.[a]	Although modeled annual average concentrations are within the Viet Nam national limits, the city aims to **meet the most stringent WHO AQG by 2030**.

[a] Enviliance ASIA. 2022. Vietnam Announces National Plan for Air Quality Management to 2025.

Source: Clean Air Asia.

Table A2.12: Ho Chi Minh City's Proposed Air Quality and Climate Actions

Priority Issues/Sector	Scope of Activities	Objective/Targets	Time Frame	Stakeholders
Industry	**Improve enforcement of industrial process emissions standards and fugitive dust control.** • Review and revise emission standards and regulations for industrial manufacturing facilities in the city.[a] • Improve inspection and compliance system. • Build the capacity of the Department of Agriculture and Environment's Environmental Permitting Unit on inspection and compliance monitoring. • Develop incentive scheme to promote adoption of cleaner technologies.	• Reduce emissions from industrial processes in facilities through enhanced emissions regulations and compliance monitoring systems.	2–5 years	• Department of Agriculture and Environment • Management Board of Industrial Parks • Department of Industry and Trade • Department of Information and Communication • Department of Education and Training • Nongovernment organizations (NGOs), civil society organizations (CSOs) • Industries/local operators
	Promote the use of alternative fuels and/or emissions abatement technologies. • Review policies and technology options for emissions abatement. • Assess the feasibility of alternative fuel applications.	• Reduced emissions from fuel combustion. • Increase the adoption of alternative fuels.	2 years	• Department of Agriculture and Environment • Management of the Board of Industrial Parks • Department of Industry and Trade • Industries and local operators

continued on next page

Table A2.12 *continued*

Priority Issues/Sector	Scope of Activities	Objective/Targets	Time Frame	Stakeholders
Transport	**Improve fuel quality and promote alternative fuels.** • Develop infrastructure to support conversion to alternative fuels (e.g., biofuel distribution system, and compressed natural gas infrastructure).	• Reduce tailpipe emissions from vehicles	>5 years	• Department of Agriculture and Environment • Department of Construction • National Government • Department of Finance
	Improve public transport infrastructure and encourage mode shift. • Develop infrastructure for public transport and expand bus network convenience, including investment in electric vehicle (EV) buses. • Develop incentive schemes and restrictions for heavy-duty vehicles in designated areas and schedules.	• Improve the public transport network and promote mode shift. • Reduce transport emissions.	2–5 years	• Department of Agriculture and Environment • Department of Construction • Department of Science and Technology • Public transport providers, businesses
	Promote the uptake of e-motorcycles. • Review and update pilot scheme for motorcycle inspection. • Install EV charging infrastructure.[a] • Assess appropriate incentive schemes to accelerate EV motorcycle adoption.	• Reduce emissions from private vehicles and motorcycles.		• Department of Agriculture and Environment • Department of Construction • Vietnam Association of Motorcycle Manufacturers (VAMM) • Electricity of Ho Chi Minh City (EVNHCMC)
Waste Management	**Implement sustainable municipal waste-to-energy initiatives.** • Develop or update the integrated waste management strategy for the city. • Update the waste-to-energy plan and complete plants currently under construction.	• Redirect 80% of waste for recycling or energy generation by 2050.	Note: Construction of waste-to-energy plants on hold as of 2022.	• Department of Agriculture and Environment • Department of Science and Technology • NGOs/CSOs

continued on next page

Table A2.12 *continued*

Priority Issues/Sector	Scope of Activities	Objective/Targets	Time Frame	Stakeholders
Agricultural Waste	**Improve management of agricultural waste.** • Develop policies to reduce agricultural waste. • Establish implementation and monitoring framework for improved practices. • Assess feasibility of alternative technologies.	• Reduce seasonal PM emissions. • Generate income from by-products for bioenergy or compost	2-5 years	• Department of Agriculture and Environment • Department of Science and Technology • Department of Finance • Department of Industry and Trade • City Police • Department of Education and Training

[a] Babela, D. et al. 2023. *Clean Air Action Plan for Ho Chi Minh City*. Ricardo.

Source: Clean Air Asia.

Iloilo City

City Air Quality Status and Emissions Profile

Iloilo City's most recent published annual average $PM_{2.5}$ levels spans from 2016 to 2017, 21 µg/m^3 and 5 µg/m^3 respectively, with 2022 at 9 µg/m^3 being the most recent recorded (footnote 1). Available records indicate that the city's annual average $PM_{2.5}$ concentrations have generally remained within the national ambient air quality standard of 25 µg/m^3. However, these levels exceed WHO AQG (5 µg/m^3) and the most stringent ITs (IT-3: 15 µg/m^3 and IT-4: 10 µg/m^3). The city's limited AQM capacity may not provide a comprehensive picture of overall air quality conditions. Efforts to assess Iloilo City's emissions profile include the development of a 2015 emissions inventory,[8] which identified mobile sources for 30% of PM (comprised of PM_{10} and ultrafine particles), 56% of NO_x, and 50% of volatile organic compounds (VOCs), largely from public utility jeepneys and motorcycles. Households contribute significantly to PM emissions due to the use of wood or charcoal for cooking.

Table A2.13: Iloilo City's Drivers and Barriers to Clean Air

	Drivers	Barriers
Policy and Regulatory Framework	• Alignment between climate and air quality action plans, with mutually reinforcing measures targeting key sectors such as transport and energy. • Cohesive policies and programs that addressing both greenhouse gas emissions and air pollution mitigation.[a]	• Risks of policy shifts at the national level affecting jeepney modernization investments by consolidated operators.

continued on next page

[8] GIZ. ASEAN-German Technical Cooperation, Clean Air for Smaller Cities in the ASEAN Region. Emission Inventory of Major Air Pollutants in Iloilo City (Final Report). Unpublished.

Table A2.13 *continued*

	Drivers	Barriers
Institutional Capacity and Coordination	• Dedicated departments addressing air quality and climate issues, with active involvement and support of different technical units.[b] • Active multistakeholder participation in air quality and climate project planning and implementation.	• Low public acceptance of certain measures, such as public transport modernization.
Data and Technology	• Trained on data governance for sustainable urban mobility planning.[c] • Demonstrated collaboration with academic institutions on research and data collection to support clean air planning.	• Reliance on external partners for technical tools and expertise. • Heavy dependence on external funding and expertise for data analytics.
Financing and Resource Mobilization	• Readiness for public–private partnership arrangements for high-impact projects, i.e., bus rapid transit and waste-to-energy. • Active regional and international partnerships supporting clean air programs and research.	• Limited data and technical knowledge for proposal development.

[a] Hechanova, N. R. 2022. *Iloilo City: Strengthening Urban Resilience & Governance In 2nd High-Level Support for Educating Sustainable Cities* (2HLS-ESC).
[b] Executive Order No. 169 Series of 2024, An Order Creating the Iloilo City Climate Change Adaptation and Mitigation Core Team (ICCAMCT) and Defining Its Functions and Responsibilities.
[c] Clean Air Asia. Iloilo City Strengthens Sustainable Urban Mobility Planning Through Training Workshop & Data.
Source: Clean Air Asia.

Table A2.14: Iloilo's City Vision for Clean Air and Climate Commitments

Clean Air and Climate Commitments	Targets
Iloilo City Local Climate Change Action Plan, 2025–2031 (unpublished draft). The city is currently updating its comprehensive climate change action plan to strengthen mitigation and adaptation strategies in alignment with broader development plans and sectoral roadmaps.	• Strengthen climate mitigation and adaptation efforts citywide. • Integrate emissions reduction across priority sectors.
Iloilo City Clean Air Plan, 2015–2017. The city's first comprehensive effort to identify major air pollution sources and outline the strategies to reduce emissions across key sectors, including transport, industry, and households. The plan articulated a vision for "A progressive city with clean air, blue skies and a community committed to sustainable development." This vision aligns with broader urban planning and climate goals.	• Identify major air pollution sources and reduce emissions from transport, industry, and households. • Promote blue skies and clean air through integrated air quality strategies. • Align air quality goals with city development plans.
BreatheLife Campaign (joined the campaign in 2018). Commitment to a multistakeholder approach focusing on (i) sustainable transport and nonmotorized transport programs, (ii) improved waste management, (iii) clean cookstoves and cleaner fuel, and (iv) strengthening stakeholders' awareness.[a]	• Implement sustainable and nonmotorized transport programs. • Improve waste management to reduce open burning and methane. • Promote clean cookstoves and cleaner fuels.
ICLEI – Local Governments for Sustainability Cities for Climate Protection Campaign. An active member since 2005 adopting a co-benefits approach to reduce both carbon and air pollutants, measures in (i) sustainable transport, (ii) energy efficiency and clean energy, and (iii) green urban planning.[b]	• Reduce greenhouse gas and air pollutant emissions through co-benefits strategies.

[a] BreatheLife. n.d. BreatheLife Welcomes Iloilo City, Philippines.
[b] Government of Iloilo City. Iloilo City Local Climate Change Action Plan, 2025–2031. Unpublished.
Source: Clean Air Asia.

Table A2.15: Iloilo City's Proposed Air Quality and Climate Actions

Priority Issues/Sector	Scope of Activities[a]	Objective/Targets	Time Frame	Stakeholders
Transport	**Implement the Metro Iloilo City bus rapid transit system.** • Establish high-density circular routes within the city. • Expand coverage to key terminals and business areas. • Extend to circumferential and metropolitan roads.	• Improve the mass public transport system.	>5 years	• City Planning and Development Office • Transport Cooperatives • Department of Transportation Region 6 • City Environment and Natural Resources Office (CENRO)
	Implement urban mobility and accessibility programs. Enhance interconnectivity and promote healthy lifestyle and inclusivity.	• Develop a 15-minute walkable city (access to basic needs within 15 minutes). • Achieve at least 20% of bicycle mode share among the public and workers	>5 years	• City Engineer's Office • CENRO • Public Order and Safety Management Office
	Develop strategies for low-carbon, renewable energy procurement and transition plans for local government unit (LGU) operations. • Transition LGU fleets to low-carbon transport vehicles (e.g., ambulances, emergency response vehicles, rubber boats). • Procure emergency vehicles compliant with Euro V emission standards.	• Achieve 100% conversion of emergency use vehicles in LGU operations by 2026 and full conversion of remaining LGU use vehicles before end of the term. • Procure at least two emergency all-weather vehicles.	1–5 years	• City Disaster Risk Reduction and Management Office
Waste Management, Energy	**Deploy waste-to-energy facilities.** • Pilot a waste-to-energy project.[a] • Conduct a feasibility study on renewable energy opportunities for the Iloilo City Sanitary Landfill. • Implement a Landfill Gas Recovery and Utilization Project in Iloilo City Sanitary Landfill. • Formulate the Iloilo City Comprehensive Circular Economy Plan.	• Design and implement climate-resilient transport, energy, and telecommunication infrastructure. • Establish one landfill gas recovery utilization project.	>5 years	• CENRO • General Services Office • Private developers and investors

[a] City Government of Iloilo City. 2025. Iloilo City Local Climate Change Action Plan 2025–2031. Unpublished.

[b] On 19 December 2023, the Iloilo City government signed a joint venture agreement under a public–private partnership scheme with MetPower Venture Partner Holdings, Incorporated, Metro Pacific Water Investments Corporation, and Metro Pacific Iloilo City Water for the construction of the first waste-to-energy project, an integrated solid waste management facility located in Brgy. Ignore, Lapaz. This facility will convert waste into refuse-derived fuel and biogas. The energy produced is to contribute to the power supply of the city's desalination plant.

Source: Clean Air Asia.

Jakarta

City Air Quality Status and Emissions Profile

The most recent available city-level air quality data for Jakarta covers 2018 and 2019 (footnote 1). In both years, the city's annual average $PM_{2.5}$ concentrations exceeded not only Indonesia's national air quality standard (15 µg/m^3), but also WHO AQG (5 µg/m^3) and all ITs (IT-1: 35 µg/m^3, IT-2: 25 µg/m^3, IT-3: 15 µg/m^3, and IT-4: 10 µg/m^3).

Figure A2.7: Jakarta's $PM_{2.5}$ Levels Against Various Air Quality Standards, 2016–2021

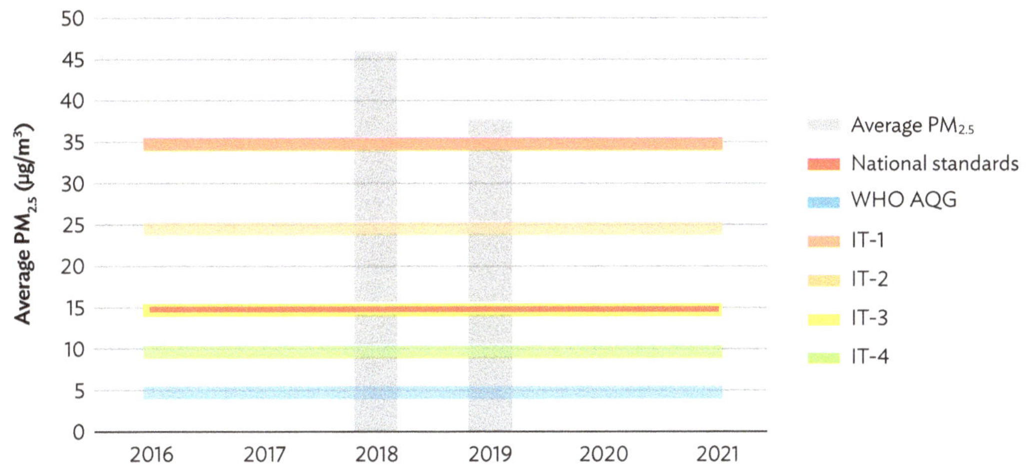

AQG = Air Quality Guidelines, IT = interim target, PM = particulate matter, WHO = World Health Organization.
Source: Clean Air Asia.

Figure A2.8: Jakarta's $PM_{2.5}$ Emissions Distribution per Sector

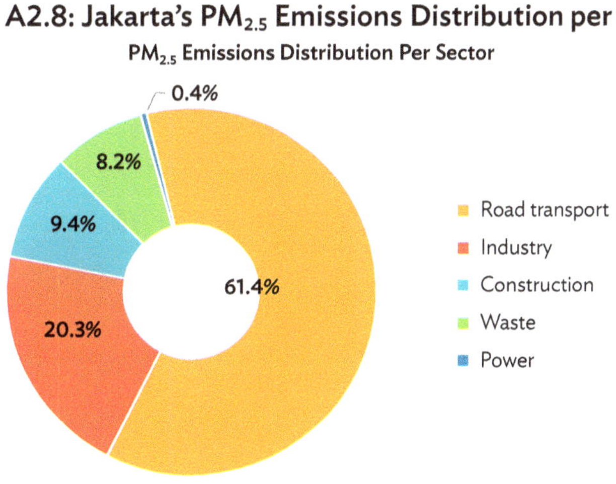

PM = particulate matter.
Source: Clean Air Asia.

In 2023, the Bandung Institute of Technology conducted a comprehensive emissions inventory for Jakarta, identifying significant $PM_{2.5}$ contributions from key sectors including transport, industry, waste, construction, and power generation.[9] Within the transport sector, emissions were further disaggregated by vehicle type: two-wheel private vehicles (2W Private), two-wheel ride-hailing services (2W Hailing), four-wheel private vehicles (4W Private), and four wheel ride hailing services (4W Hailing).

[9] Systemiq & Bandung Institute of Technology. 2025. Better Air, Better Indonesia: Executive Summary Systemiq.

Table A2.16: Jakarta's Drivers and Barriers to Clean Air

	Drivers	Barriers
Policy and Regulatory Framework	• A range of local policies and measures addressing emissions from transport and industry and urban green space development.[a] • An action plan targeting major sources of air pollution developed with stakeholder support. • National targets and government incentives to support EV adoption.	• Limited sense of urgency and low consumer awareness. • Weak enforcement of policies and regulations. • Regulatory gaps in other sectors (waste, industry, etc.).
Institutional Capacity and Coordination	• Active multistakeholder participation in air quality and climate project planning and implementation. • An established Air Pollution Taskforce at the provincial government level.	• Limited coordination between local Jakarta provincial government and national ministries, and between Jakarta Environmental Agency and other local governments in the Greater Jakarta area. • Adjustment of distribution logistics, especially in the importation of Euro IV compliant fuels.
Data and Technology	• Government-led digital application to manage access to fuel subsidies (i.e., MyPertamina application). • Digitalized monitoring and reporting systems including mandatory electronic reporting for environmental impact assessments and integration of Continuous Emissions Monitoring System (CEMS) data into Continuous Industrial Emissions Monitoring Information System (SISPEK).	• Limited charging and battery-swapping infrastructure. • Need to upgrade electrical grids to meet higher power demand.
Financing and Resource Mobilization	• Potential growth of the EV market and associated government savings from fuel subsidy reforms.	• High upfront investment cost: $4.1 billion estimated government spending to support key interventions i.e., public Euro IV implementation, transport expansion, open waste burning mitigation, and construction dust control

EV = electric vehicle.

[a] Governor of DKI Jakarta Instruction No. 66.2019 on Air Pollution Control.

Source: Clean Air Asia.

Table A2.17: Jakarta's City Vision for Clean Air and Climate Commitments

Clean Air and Climate Commitments	Targets
Breathe Cities. Jakarta is a participating city in this global initiative, led by Bloomberg Philanthropies, the Clean Air Fund, and C40 Cities. This initiative aims to combat air pollution and climate change in urban areas.	Reduce air pollution by up to 30% by 2030.
Cities4Forests.[a] A collaboration with World Resources Institute (WRI) Indonesia to advance support for nature-based solutions for flood mitigation and support GHG inventory development.	50% emissions reduction by 2030.

continued on next page

Table A2.17 *continued*

Clean Air and Climate Commitments	Targets
Global Covenant of Mayors for Climate and Energy (GCoM).[b] GCoM is the largest global alliance for city climate leadership. In 2016, Jakarta became a signatory of GCoM, committing to reduce GHG emissions, strengthen climate risk adaptation, and expand access to clean and affordable energy.	Reduce GHG emissions, identify and adapt to climate change risks, and increase access to clean and affordable energy.

GHG = greenhouse gas.

[a] World Resources Institute Indonesia. 2024. Cities4Forests synergy with the DKI Jakarta Provincial Government Strengthens the Commitment to Nature-Based Climate Action for Jakarta.

[b] Asian Mayors. 2025. DKI Jakarta and Pontianak, Two Indonesia's GCoM Signatory Cities Recognized for Advancing Climate Action.

Source: Clean Air Asia.

Table A2.18: Jakarta's Proposed Air Quality and Climate Actions

Priority Issues/Sector	Scope of Activities	Objective/Targets	Time Frame	Stakeholders
Transport Emissions	**Develop integrated public transport systems.** • Expand MRT and LRT routes. • Strengthen physical and service integration across networks.	• Establish routes and infrastructure that is fully integrated with existing public transport systems.	>5 years	• Provincial Government of Jakarta • Ministry of Transportation • Ministry of Finance
	Accelerate adoption of e-vehicles (E2W, E4W, e-buses, e-trucks) and expand city-wide charging infrastructure.	Increase the share of EVs in Jakarta: • E2W: 45% (5.5 million) • E4W: 23% (887,000) • E-trucks: 15% (115,000) • E-buses: 100% (10,000)	>5 years	• E2W/E4W/E-bus/E-truck and battery OEMs • Local government agencies • State Electricity Company (PLN) • Ministry of Energy and Mineral Resources
Waste Management	**Eliminate open waste burning.** Explore long-term waste management solutions including waste treatment and energy recovery (e.g., gasification and pyrolysis), upgrades to sanitary landfills, and Extended Producer Responsibility.	• Eliminate open waste burning.	>5 years	• Jakarta Environmental Agency • Ministry of Environment

continued on next page

Table A2.18 *continued*

Priority Issues/Sector	Scope of Activities	Objective/Targets	Time Frame	Stakeholders
Industry and Energy Emissions	**Upgrade or retire CFPs**	• Achieve 100% CFP compliance with to post-2019 emission standards.	>5 years	• Ministry of Environment • Ministry of Finance • Ministry of Industry
	Achieve cleaner technologies for industrial boiler. Adopt best-available technology for e-heating solutions as low-carbon and low-pollutant alternatives.	• Achieve 0% coal usage in industrial boilers.	>5 years	• Ministry of Environment • Ministry of Finance • Ministry of Industry

EV = electric vehicle, CFPs = coal-fired power plants, E2W = electric two-wheelers, E4W = electric four-wheelers, LRT = light rail transit, MRT = mass rapid transit.

Source: Systemiq and Bandung Institute of Technology. 2025. Better Air, Better Indonesia—The Economic and Political Case for Urgent and Coordinated Action for Jakarta's Clean Air.

Phnom Penh

City Air Quality Status and Emissions Profile

Phnom Penh's most recent city-level air quality data covers 2017 to 2023 (footnote 1). During this period, annual average $PM_{2.5}$ concentrations remained within the range of the national standard (50 µg/m^3) but exceeded WHO AQG (5 µg/m^3) and all ITs (IT-1: 35 µg/m^3, IT-2: 25 µg/m^3, IT-3: 15 µg/m^3, and IT-4: 10 µg/m^3).

Figure A2.9: Phnom Penh's $PM_{2.5}$ Levels Against Various Air Quality Standards

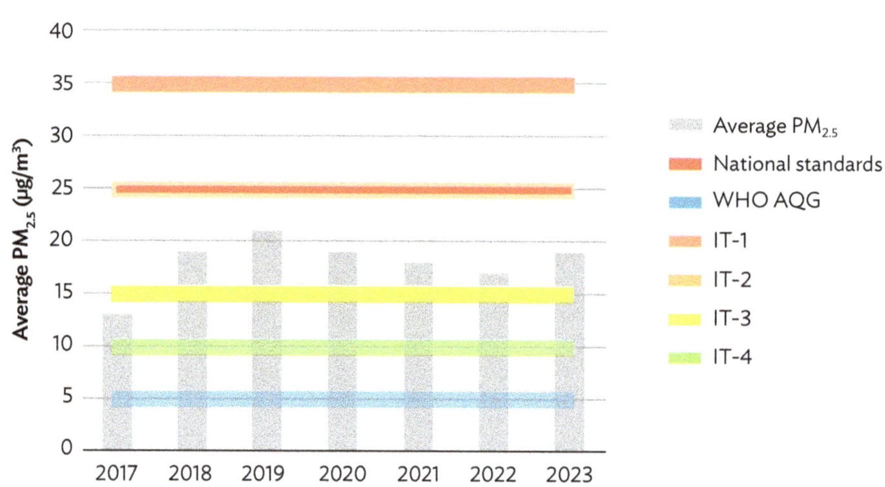

AQG = Air Quality Guidelines, IT = interim target, PM = particulate matter, WHO = World Health Organization.

Source: Clean Air Asia.

The key drivers of air pollution in the city, based on the emissions profile of Phnom Penh from the 2019 National GHG Emissions Inventory Report[10] and Clean Air Plan of Cambodia,[11] are presented in Figure A2.10.

Figure A2.10: Phnom Penh's $PM_{2.5}$ Emissions Distribution per Sector

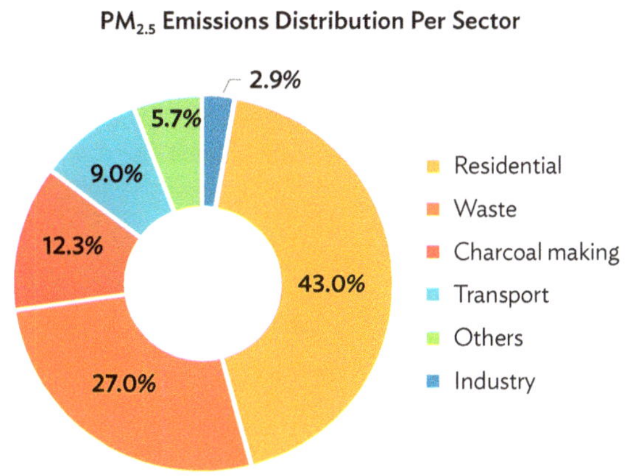

PM = particulate matter.
Source: Clean Air Asia.

Figure A2.10 indicates that residential fuel use is the largest contributor to $PM_{2.5}$ emissions (43%), followed by waste burning (27%) and charcoal production (12.3%). While charcoal production is more common in rural areas, small-scale producers also operate in Phnom Penh, particularly in peri-urban zones, contributing to local air pollution.

Table A2.19: Phnom Penh's Drivers and Barriers to Clean Air

	Drivers	Barriers
Policy and Regulatory Framework	• National framework and strategic plans addressing reducing ambient air pollution, climate vulnerability, and city-level sectoral development plan.	• Gaps in enforcement of policies and regulations.
Institutional Capacity and Coordination	• Established technical working group for the development and implementation of air quality interventions. • Partnership strategy involving multistakeholder engagement (established under the UNDP Asian Air Quality Management Program (AAQMaP).	• Limited technical capacity within government institutions for air quality management, with continued reliance on development agencies for capacity building.[a] • Limited coordination between ministries.[b]
Data and Technology	• Opportunities to expand the AQM system through the CASA project (supported by ASEAN-ROK Cooperation Fund).[c]	• Limited data on pollutant sources, emissions and levels of air quality to inform mitigation measures. Public skepticism regarding air quality data due to insufficient information disclosure.

continued on next page

10 Kingdom of Cambodia, National Council for Sustainable Development and United Nations Framework Convention on Climate Change. 2019. National Greenhouse Gas Inventory of the Kingdom of Cambodia (1994–2016).

11 Climate and Clean Air Coalition. Clean Air Plan of Cambodia.

Table A2.19 *continued*

	Drivers	Barriers
Financing and Resource Mobilization	• Some capacity building and multi-stakeholder engagement in identifying collaborative projects and financing opportunities undertaken.[d]	• Changes to national or local priorities with regards to clean air or climate action plan implementation.

[a] Global Green Growth Institute. 2024. Cambodia Government Officials Dive into Air Quality Assurance and Control at AFD-Funded Capacity-Building Workshop.
[b] United Nations Economic and Social Commission for Asia and the Pacific and Ministry of Environment, Cambodia. n.d. Air Quality Management in Cambodia.
[c] ASEAN–Korea Cooperation Fund. 2025. Handover Ceremony of Air Quality Monitoring System to Cambodia.
[d] United Nations Economic and Social Commission for Asia and the Pacific. 2024. National Workshop on Unlocking Finance to Support Clean Air Solutions in Cambodia.
Source: Clean Air Asia.

Table A2.20: Phnom Penh's City Vision Clean Air and Climate Commitments

Clean Air and Climate Commitments	Targets
Cambodia Clean Air Action Plan (2021). National Action Plan to reduce open waste burning.	• Reduce open waste burning by 50% across the country by 2030.

Source: Climate & Clean Air Coalition. 2024. Cambodia—Deliver Training Program for the Implementation of Decrees for Reducing Open Waste Burning and Air Pollution (Project KH-24-007).

Table A2.21: Phnom Penh's Proposed Air Quality and Climate Actions

Priority Issues/Sector	Scope of Activities	Objective/Targets	Time Frame	Stakeholders
Transport	**Develop or improve public transport strategy and route systems** (identified under Clean Air Plan for Cambodia). • Implement the Phnom Penh Declaration on Sustainable Urban Mobility in Cambodia.[a]	• Strengthen and improve the efficiency of the public transit system by 2030.	>5 years	• Ministry of Environment • Ministry of Public Works and Transport • Inter-Ministerial Technical Committee for the Development of EVs
	Employ EVs and develop and electric transport network in the city.[b] • Establish a scrappage scheme for ICE-3Ws *(tuk-tuks)*.[c] • Conduct a feasibility study to convert the government fleet to electric vehicles.	• Implement pilot projects in Phnom Penh.	>5 years	• Ministry of Public Works and Transport

continued on next page

Table A2.21 *continued*

Priority Issues/Sector	Scope of Activities	Objective/Targets	Time Frame	Stakeholders
Agricultural Emission and Waste Management	**Develop waste-to-energy facility** (identified under Clean Air Plan for Cambodia).[d] • Promote non-burn alternatives for agricultural waste. • Develop technical guidelines and training programs.[e]	• Install infrastructure by 2030. • Reduce waste volume and enhance recovery of waste as alternative fuel. • Reduce emissions from agricultural waste burning.	>5 years	• Ministry of Environment • Provincial Department of Environment
Industry Emissions	**Improve efficiency of charcoal production** (identified under Clean Air Plan for Cambodia). • Promote the use of cleaner charcoal alternatives.[f] • Roll out training on Resource Efficient and Cleaner Production program.	• Increase charcoal kiln efficiency • Improve worker understanding of air quality issues and enhance factory productivity.	>5 years	• Ministry of Environment • Department of Air Quality Management • Ministry of Industry, Sciences, Technology, and Innovation

[a] Gesellschaft für Internationale Zusammenarbeit (GIZ) and GFA Consulting Group GmbH. 2020. Sustainable Design of Urban Mobility in Medium-Sized Metropolitan Regions (SMMR).
[b] Based on the Clean Air Plan for Cambodia, and in combination with the implementation of the National Electric Vehicles Policy, 2024–2030 (adopted in May 2024), which focuses on promoting the growth of the electric vehicle industry.
[c] World Bank. 2024. Cambodia: Recommendations to the National Roadmap for Electric Mobility Transition (Report No. 099031924073537111).
[d] In mid-February 2021, the Urban Garbage and Solid Waste Management Committee was established to fulfill the objective that the solid waste is properly managed from the places of generation until waste processing that will turn the waste into electricity.
[e] Climate and Clean Air Coalition. 2024. Cambodia—Deliver Training Program for the Implementation of Decrees for Reducing Open Waste Burning and Air Pollution (Project KH-24-007).
[f] Phurik-Callebaut, R., R. Van Berkel, and T. Thomas. 2024. Khmer Green Charcoal: Char Briquettes from Coconut Shells. SWITCH-Asia Programme.

Source: Clean Air Asia.

Quezon City

City Air Quality Status and Emissions Profile

Quezon City's most recent city-level air quality data cover the period 2022–2024 (footnote 1). During this period, the city's annual average $PM_{2.5}$ concentrations remained within the national ambient air quality standard (25 µg/m^3). However, levels have consistently exceeded WHO AQG (5 µg/m^3) and most interim targets (ITs), including IT-4 (10 µg/m^3), IT-3 (15 µg/m^3), and IT-2 (25 µg/m^3). $PM_{2.5}$ concentrations showed a slight decline in 2024 (19.17 µg/m^3) compared to 2023, though overall levels remain well above global health benchmarks.

Figure A2.11: Quezon City's $PM_{2.5}$ Levels Against Various Air Quality Standards, 2022–2024

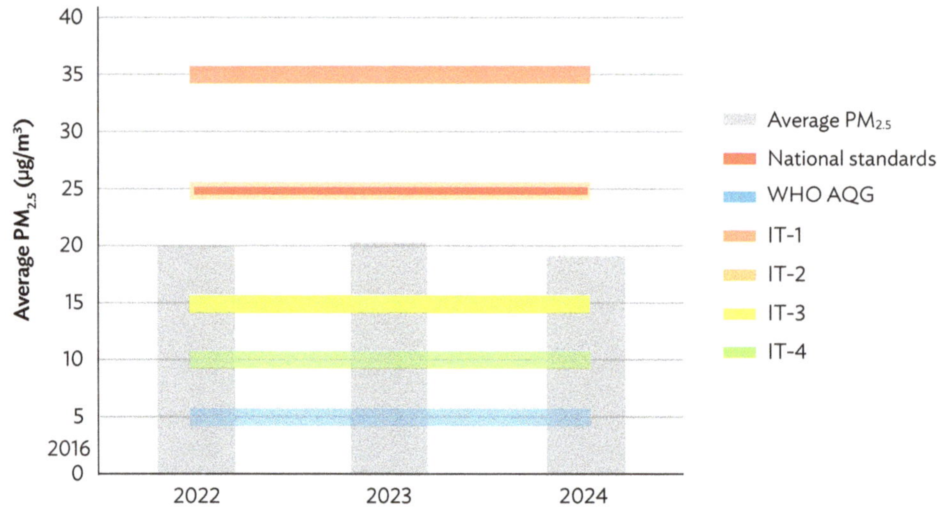

AQG = Air Quality Guidelines, IT = interim target, PM = particulate matter, WHO = World Health Organization.
Source: Clean Air Asia.

Quezon City's 2021 emissions inventory identified the transport sector as the dominant source of almost all air pollutants, with pre-Euro technology public utility vehicles as key contributors to $PM_{2.5}$ and black carbon emissions. Industrial sources, including furnaces and coal-fired boilers in manufacturing plants operating almost year-round, were also significant contributors to these pollutants (Figure A2.12).

Figure A2.12: Quezon City's $PM_{2.5}$ Emissions Distribution per Source Type, 2021

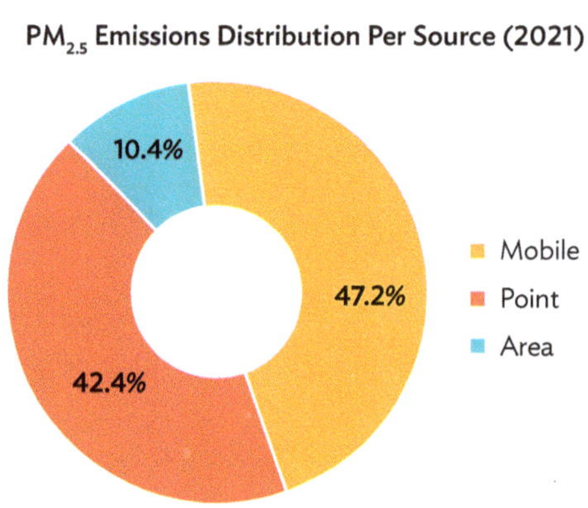

PM = particulate matter.
Source: Clean Air Asia.

Table A2.22: Quezon City's Drivers and Barriers to Clean Air

	Drivers	Barriers
Policy and Regulatory Framework	• Alignment between local climate and air quality action plans, with mutually reinforcing measures targeting key sectors, i.e., energy, transport, and waste. • Policy coherence across local, national, and regional development frameworks and goals.	• Gaps in enforcement and monitoring of existing policies. • Absence of specific local policy and regulations to enforce or operationalize green zones or routes.
Institutional Capacity and Coordination	• Dedicated department for air quality and climate issues—Climate Change and Environmental Sustainability Department, with trained technical staff.[a] • Established Air Quality Technical Working Group to coordinate across departments/units and align with city development plans. • Active multistakeholder participation in air quality and climate planning and implementation.	• Limited technical capacity in data management, analysis, and application of air quality information in local planning (city and barangay levels). • Limited integration of monitoring results into program and activity planning.
Data and Technology	• Established AQM network within the city. • Use of digital tools and platforms for air quality reporting.	• Limited integration of air quality data into target setting and planning. • No long-term or institutionalized funding for air quality sensors and experts.
Financing and Resource Mobilization	• Availability of local funds and donor programs for air quality and climate-related initiatives. • Active leadership in mobilizing funding and investment opportunities.	• Capacity-building and monitoring activities remain largely supported by donor-driven projects.

AQM = air quality management.

[a] List of capacity-building activities that were identified in Table 94 of the Quezon City Air Quality Management Plan.

Source: Quezon City Government. 2023. *Quezon City Air Quality Management Plan* and Clean Air Asia.

Table A2.23: Quezon City's Vision for Clean Air and Climate Commitments

Clean Air and Climate Commitments	Targets
Quezon City AQMP.[a] Released in 2023, the AQMP provides a comprehensive roadmap to reduce emissions of key air pollutants. Through this plan, the city aims to • Adopt a city-wide air quality and emissions reduction strategy aligned with WHO guidelines, national standards, and Quezon City's enhanced local climate change action plan (eLCCAP). • Implement targeted measures across transport, waste, energy, industry, and construction sectors. • Monitor and evaluate progress through data-driven assessments, institutional coordination, and integration with the city's broader climate and development plans.	Emission reduction targets range from 25% to 87% by 2040, depending on the pollutant ($PM_{2.5}$, SO_2, NO_x, BC, OC, VOCs, CH_4, CO).
BreatheLife Campaign.[b] Quezon City joined the BreatheLife campaign in 2020 as part of a global effort led by WHO, UNEP, and Climate and Clean Air Coalition. The city committed to strengthen its clean air actions and work toward achieving WHO AQG by 2030.	Aim to meet WHO AQG by 2030 as part of a long-term strategy spanning 2020–2040.

continued on next page

Table A2.23 *continued*

Clean Air and Climate Commitments	Targets
Clean Air Cities Declaration (C40).[c] In 2019, Quezon City signed the Clean Air Cities Declaration during its engagement with the C40 Cities Climate Leadership Group. Through this commitment, the city pledged to: • Set ambitious air quality targets to meet or exceed national standards or the WHO guidelines. • Implement substantial new policies and programs to address key sources of air pollution within the city. • Establish baseline air pollution levels and develop implementation plans that outline actions and milestones to meet their air quality goals.	• Adopt WHO-aligned air quality targets that started in 2019.

AQG = Air Quality Guideline, AQMP = Air quality management plan, BC = black carbon, CH_4 = methane, CO = carbon monoxide, NO_x = nitrogen oxides, OC = organic carbon, PM = particulate matter, SO_2 = sulfur dioxide, UNEP = United Nations Environment Programme, VOCs = volatile organic compounds, WHO = World Health Organization.

[a] Clean Air Asia. 2024. Draft Quezon City Air Quality Management Plan.

[b] BreatheLife. 2020. Quezon City, Philippines Joins BreatheLife Campaign.

[c] C40 Cities Climate Leadership Group. 2019. Clean Air Cities Declaration: Quezon City Joins Global Cities in Committing to Clean Air.

Source: Clean Air Asia.

Table A2.24: Quezon City's Proposed Air Quality and Climate Action

Priority Issues/Sector	Scope of Activities	Objective/Targets	Time Frame	Stakeholders
Transport	**Implement green zones to reduce emissions from mobile sources and improve public space.** • Select pilot green zones. • Conduct baseline air quality and emission assessments. • Develop local policies formalizing green zones and routes. • Develop plan for expansion and replication based on results.	• Improve air quality and urban environment. • Reclaim public space for sustainable mobility.	>1 year	• TTMD • DPOS-GTD • CCESD
	Promote public or mass transport. • Augment city bus services, including route expansion.	• Provide viable transport options for commuters and reduce reliance on private vehicles. • Reduce GHG and pollutant emissions from transport.	>15 years	• TTMD • Office of the City Mayor • Office of the City Administrator

continued on next page

Table A2.24 *continued*

<table>
<tr><th>Priority Issues/Sector</th><th>Scope of Activities</th><th>Objective/Targets</th><th>Timeframe</th><th>Stakeholders</th></tr>
<tr><td rowspan="2">Transport</td><td>Promote fleet electrification and introduce vehicle management systems.
• Transition the city vehicle fleet to EVs.[a]
• Conduct needs assessment for fleet management and maintenance, and infrastructure and technology.
• Build capacity for EV fleet management within city departments.</td><td>• Increase share of EVs in the city government fleet.
• Reduce GHG and pollutant emissions in government facilities.</td><td>In progress 2023–2026</td><td>• City General Services Department
• DPOS-GTD</td></tr>
<tr><td>Encourage active school transport (Bike/Walk to School).
• Develop bicycle and pedestrian network plans.
• Implement awareness and safety programs.
• Collaborate with school communities for implementation.</td><td>• Improve air quality near schools and along bike corridors.
• Increase the number of students biking and walking.</td><td>In progress 2024–2026</td><td>• DPOS-GTD
• Task Force on Bicycle and Active Transport
• Schools Division Office of Quezon City
• School Administration</td></tr>
<tr><td rowspan="2">Waste Management</td><td>Establish a city-wide community composting for biodegradable waste.
• Set up composting hubs at barangay level.
• Conduct training and awareness campaigns.</td><td>• Increase the volume of biodegradable waste diverted for composting.
• Increase the number of barangays equipped with composting hubs.</td><td>3 years</td><td>• CCESD
• DSQC
• Solid Waste Management Board
• Market Development and Administration Department</td></tr>
<tr><td>Establish community composting for biodegradable waste: anaerobic biodigesters in public markets.
• Identify pilot sites (public markets) for installation of biodigesters.
• Train personnel for operation and maintenance.</td><td>• Reduce methane emissions from organic waste.</td><td>>5 years</td><td>• DSQC
• City Health Department
• CCESD</td></tr>
</table>

CCESD = Climate Change and Environmental Sustainability Department, DPOS-GTD = Department of Public Order and Safety – Green Transport Division, DSQC = Department of Sanitation and Cleanup Works of Quezon City, EVs = electric vehicle, GHG = greenhouse gas, IBAQ = Integrated Program for Better Air Quality, TTMD = Traffic and Transport Management Department.

[a] A cost-benefit analysis was carried out that identified at least 14 years old as priority units for replacement. This study was developed with support from the Clean Air Asia under the IBAQ Program.

Source: Clean Air Asia.

www.ingramcontent.com/pod-product-compliance
Lightning Source LLC
Chambersburg PA
CBHW041757220926
51847CB00014B/235
* 9 7 8 9 2 9 2 7 7 7 4 8 7 *